I0482736

NIST Technical Note 1622

Clean Agent Suppression of Energized Electrical Equipment Fires

Gregory T. Linteris
Fire Research Division
Building and Fire Research Laboratory
National Institute of Standards and Technology

January 2009

U.S. Department of Commerce
Carlos M. Gutierrez, Secretary

National Institute of Standards and Technology
Patrick D. Gallagher, Deputy Director

Executive Summary

The NFPA 2001 standard on the use of clean agents for the suppression of fires arose from the phase-out of Halon 1301. Standard methods exists for specifying the amount of clean agent required for Class A and Class B fires, but the recommendation for Class C fires (those involving energized electrical equipment) defaults to the Class A values. While this may be appropriate for some Class C fires, there is concern that higher agent concentration may be necessary if energy is added to the fire by the electrical source. A number of test methods have been proposed to determine the amount of agent required to suppress fires in energized electrical equipment; however, there has been no broad agreement on a test method to include in NFPA 2001 for Class C fires. Further, some of the test methods suggested that the current recommended total flooding concentration is sufficient, while others suggested that higher concentrations may be necessary for some fires. The present project was sponsored by the Fire Protection Research Foundation to address the need for a standard test to be included in NFPA 2001 for Class C fires. The goals of the project are to understand the fire threats occurring in energized electrical equipment, and suggest a test protocol which can properly estimate the amount of agent required to suppress fires in those situations.

As a first step, phone interviews were conducted with members of the technical panel and with the sponsors of the present project. These representatives, as well as other expert sources recommended by them, provided information on the likely fire threats expected in the field. Detailed case studies were supplied by FM Global. Notes of the conversations as well as summaries are provided, and these served as one source for definition of the threat.

In order to illustrate the relevant parameters necessary to consider in fires over solid materials with added energy, a thorough literature review was performed. Topics included materials flammability and fire suppression, with the latter broken down into: a theoretical description of fire suppression, flow-field effects, effects of heat addition on suppression, and suppression of flames over condensed-phase materials.

A major resource was previous published work to develop standard tests for suppression of fires in energized electrical equipment. By analyzing these in detail, and considering the relevant physics of the suppression processes, the desired properties of a standard test were developed, and the range of values of the most important parameter (the flux of added energy) was estimated. Major findings of the report are listed below.

1. Added energy to a burning material will affect the minimum extinguishing concentration of suppressant, and the effect is relatively independent of the energy source (e.g., resistive heating, radiative heating, conduction, preheating of the material, etc.) as long as the energy is added to the burning surface.

2. For most of the tests which have been proposed to simulate fires in electrical equipment, the magnitude of the estimated external heat flux is similar to that which can be obtained with radiant heating experiments.

3. In the test methods which have been proposed to simulate the suppression of flames in energized electrical equipment, the required suppressant concentration in the different tests agreed with each other reasonably well at equivalent energy input levels (when enough information was provided to do a comparison).

4. Many of the methods previously proposed do not combine the relevant parameters in ways which produce the most conservative (yet plausible) test.

5. A test method based on an external radiant heat flux source, a large sample (10 cm x 10 cm) in a chimney, and realistic materials is proposed.

6. Two radiant flux levels in the proposed test method are suggested:
   a.) an incident flux of 20 kW/m^2 as a lower limit, representing the heat input without any electrical augmentation, but with an adjacent flame on similar burning material (which enhances burning);
   b.) an incident flux of 50 kW/m^2 for cases representing sufficient electrical energy to bring the polymer to its decomposition temperature and maintain it there (in the absence of the flame). An example of this would be an energized cable fire.

7. To assign appropriate energy flux levels for electrical power addition intermediate between these two limits, better understanding of specific electrical failure modes is required.

8. An approach to determine the realistic power levels for situations between the limiting cases has been suggested. The first steps are to survey the fire suppression industry and to collate statistical data on electrical fire incidents. These must be followed, however, by three additional steps: obtaining input from electrical equipment hardware experts (or experts in forensic investigation of electrical failures), performing laboratory experiments, and modeling to simulate the likely failure events so that the values of the relevant controlling parameters can be estimated.

Table of Contents

List of Figures

List of Tables

1. Introduction

1.1 Problem Description

The suppression of fires by clean agents (those that leave no residue and are not electrically conductive) is covered by the National Fire Protection Association (NFPA) Standard on Clean Agent Fire Suppression Systems, NFPA 2001 [1]. This standard was developed in response to the phase-out of the effective and widely used agent Halon 1301. Fires are typically classified as Class A, B, C, D, or K, and NFPA 2001 describes test procedures to be used in determining the design extinguishing concentration of the agents based on the fire class. For Class A fires, a testing procedure is required which meets, at the minimum, the procedures in UL 2127[*] (for inert agents) or UL 2166 (for clean agents, typically hydrofluorocarbon, HFC; or hydrochlorofluorocarbon, HCFC agents). The minimum design concentration of the agent is that determined in the UL test times a safety factor of 1.2. For Class B fires, the cup burner test is specified, and the minimum design concentration is the cup burner value times a safety factor of 1.3. (Class D fires are not covered, and Class K is a subset of Class B). Fires involving energized electrical equipment (Class C) are covered in Section 5.4.2.5 of NFPA 2001, which states:

> "Minimum design concentration for Class C hazards shall be at least that for Class A surface fire."

While it is desired to remove the power from the electrical equipment prior to fire suppression, that decision can be at the discretion of the equipment owner, taking into consideration 1.) ancillary loss of life due to the shutdown, 2.) fire threat to occupants or property, 3.) economic loss due to loss of function, and 4.) economic loss due to facility damage. Hence, there are cases where fire suppression systems will be designed under the assumption that energized electrical equipment will be present. Unfortunately, there exists no standard test method for the amount of agent necessary to suppress fires in cases in where the combustion may be augmented due the addition of energy from an electrical source.

1.2 Background

The problem of fire protection in electrically energized environments has been discussed in review articles [2,3], and several test methods have been used to simulate the effects of energized electrical equipment. These include tests which strive to suppress a flame over a realistic electrical failure event, with representative polymeric materials [4-7], and those that attempt to control the salient parameter (the external heat flux, EHF) [8-11]. In general, the test results have shown that higher agent concentration is required in the presence of energy input from the different sources [8-13], while the results in ref. [4-7] indicated that a typical design concentration of 7 % (by volume) for HFC-227 in Class A fires was sufficient to extinguish their test cases (although modification of one of the tests and the extinguishment criterion in ref. [5] implied a higher agent concentrations for suppression [14,15]). Despite the extensive work, no generally accepted test procedure has emerged, and no consensus exists as to the relationship between any of the test methods and the actual fire threats.

---

[*] A portion of this work was carried out by the National Institute of Standards and Technology (NIST), an agency of the U. S. government and by statute is not subject to copyright in the United States. Certain commercial equipment, instruments, materials or companies are identified in this paper in order to adequately specify the experimental procedure. This in no way implies endorsement or recommendation by NIST. The policy of NIST is to use metric units of measurement in all its publications, and to provide statements of uncertainty for all original measurements. In this document however, data from organizations outside NIST are shown, which may include measurements in non-metric units or measurements without uncertainty statements.

1.3 Objective

There are two main objectives of the present work: 1.) to define typical fire hazards for applications involving energized electrical equipment (Class C fires) that normally are protected by clean agent fire extinguishing systems, and 2.) to develop and suggest a test protocol that can provide scientifically justified minimum extinguishing concentrations of clean agents required to protect typical energized electrical equipment.

1.4 Approach

The first step in this work was to assemble information and input from the technical panel members and corporate sponsor representatives of the NFPA Fire Protection Research Foundation Project on Clean Agent Suppression of Energized Electrical Equipment. Following that, the literature dealing with suppression of burning polymers was reviewed. Topics included fire suppression, material flammability, suppression of flames over condensed-phase materials, and finally, the effect of energy addition on the suppression of flames over materials. The focus of the reviews was information essential to interpretation of both the actual suppression of fires in energized electrical equipment and any test designed to mimic the fire suppression. Based on the information provided by the technical panel (and the sources which they recommended) we attempted to define the threat to be controlled by the clean agent systems. With those threats in mind, the existing tests and data were reviewed extensively. Many of the tests proposed were based on the principle of replicating, in the laboratory, the actual threats expected in the field. Hence, they could serve as useful test cases for which more detailed analysis could be performed. From the interpretation of the phenomena occurring in the test methods, more insight could be gained concerning the phenomena as well as the tests. Using the analyses of the test methods and their results, we generate a list of desirable properties in a standard test, and then recommend a test method based on that list. Finally, we recommend further research which would help to better define the threat, which would allow a better specification of the test method.

2. Results

2.1 Information Assembly: Phone Interviews with Technical Experts

2.1.1 Overview

The goal of this section of the work plan was to assemble information to define the fire threat to be suppressed by clean agents in electrically energized equipment fires. To do that, information on critical applications, equipment types, fire threats, potential clean agent applications, agent discharges, and reported incidences is required. Ultimately, a thorough survey of facility owners, agent suppliers and equipment installers is needed; however, this is beyond the scope and time scale of the present project. As a first step in understanding the problem, we gathered input from experts who comprise the technical panel and the sponsors for the NFPA Research Foundation Clean Agent Suppression of Energized Electrical Equipment project. The goal was to elicit from them their perspective on the project, as well as to obtain from them data on actual suppressed energized electrical equipment fires, or names of contacts who may have such information. Phone interviews were conducted with all but one of the technical panel, and experts from all but two from the sponsor organizations, as well as several individuals who are world-renowned experts in the fields of materials flammability and ignition. The conversations were very interesting and informative, and notes from the phone conversations are listed in Appendix I.

In general, most (but not all) of the interviewees either did not have (or did not want to discuss) specific fire situations in any detail. This is understandable, and illustrates the general difficulty of defining the fire threat to be suppressed by clean agents. It seems a reasonable policy to insure that any information publicly released is both accurate and protects the identity of individuals involved, and doing this requires significant effort to prepare the materials for release. It may also be that the organizations themselves do not have available the details of what happened in the fires, especially at the level of detail necessary to simulate and replicate what happened (which is essentially what a test method tries to do). Several of the respondents did give detailed accounts of some fires, and FM Global provided written descriptions of three case histories (which are provided in Appendix II).

2.1.2 Results of Phone Survey of Project Technical Panel and Sponsors

The phone interviews provided excellent background to the problem. The responders represented national or international experts on the topic, with a wealth of practical information. Their responses are organized below with regard to specific topics.

<u>Need for Addressing Effects of Energy Augmentation on Fire Suppression</u>

Some of the respondents felt that there is no problem, and that the topic was really a non-issue, while others felt that a major fire in a data processing center (DPC) or telecom central office (TCO) is a disaster waiting to happen—that it's inevitable. The former respondents cited the small number of fires in telecom and data processing which have occurred so far, and the highly successful fire prevention rate. The latter felt that the good record so far was due largely to the success of the Network Equipment Buildings Systems (NEBS) used in telecom. They believed that the new buildings housing telecommunications and data processing equipment are much more varied, do not follow a single standard as stringent as NEBS, and hence are more vulnerable than such buildings have been in the past.

Most of the respondents in between these two extreme views felt that there probably are differences in behavior when suppressing electrically energized equipment fires, and that it's best to do the right thing and try to understand them, and incorporate that understanding into a test. That is, there probably are some electrically energized equipment fires that will not be put out by design concentrations resulting from the current NFPA 2001 tests, and it would be good to be able to understand what those conditions are.

Three of the respondents had a similar view: that the problem was ill-defined. Electrical ignition sources are just that: ignition sources. After the ignition occurs, there is little energy addition from the ignition source, and the fire moves on to another location. Two of these three, however, felt that the nature of some electrical ignitions is such that they create a much larger initial ignition site. Since the fire is much larger from the outset, the usual arguments about radiant heating from adjacent flames applies, and one has to test for the material burning and suppression with added radiant heat loads typical of larger fires.

Everyone who mentioned it agreed that in electrically energized equipment fires, if the power is left on, there is a likely possibility of re-light after the suppressant concentration decays. There was always a general acceptance that if energy is added to the system, the quantity of agent required is higher. Several respondents noted that there was no situation in their facilities in which a clean agent would be released into an electrically energized equipment fire.

Need for Better Information on Actual Fire Threats to be Suppressed by Clean Agents

Nearly everyone, except some of those who felt that it was a non-issue, felt that there was a need for better information on what the fire threats actually are.

Likelihood of Power-down with a Fire

Of the respondents who discussed it, there was almost unanimous feeling that in telecommunications central offices or data processing centers, everyone is trained to avoid shutdown, and it was very unlikely that employees would shut down the facility in the event of a fire. The sentiment was that intentional shut down would only occur if there were no other choice—or perhaps even never at all. On the other hand, some said that while shut down was very unlikely, their *policy* was to shut down before releasing agent.

Value of Central Power-Down Switch or Procedure

Many felt that in the event of a localized fire, the problem of de-powering would be much easier if there were a single-point shutdown switch, or at least a well-specified shutdown procedure. But many also felt that a single-point shutdown was either not practical (the systems were too complicated), or that such a switch made system failure more likely (due to mistakes, single-point failure, or sabotage). Better procedures and training for shut down were generally agreed upon by those discussing it.

Relevant Size of Electrical Sources of Energy, Power Levels to Consider

Nearly everyone felt that the problem was very broad, with a very large range of electrical energy input possible. Nearly everyone also felt that the problem could be limited to telecommunications central offices and data processing centers, since those represented 80 % to 95 % of the clean agent system installations in the field. There was a general agreement from most that, even in these situations, power cables should be treated differently from data cables. Several respondents noted that power cables are sometimes un-fused, and the over-current devices can fail, so in some systems the power can be limited only by the cable size (typically oversized to limit voltage drop), and the current capacity of the battery back-up system. Hence, power levels of up to 4000 kW are possible. On the other hand, for data lines, power will be limited to a few hundred watts. A few respondents felt that the power going into a cabinet (typically 1500 W) was the limiting power, and this should be one category (separate from power cables). One respondent suggested that the power level of the ignition fire in the NEBS rack-level test (average value of 2.5 kW, peak 5 kW) was an appropriate level of power to consider. Many respondents felt that for energy input above a certain amount, clean agents (at the levels at which they are typically added) won't put out the fire, so that feature should be brought out and made clear in the literature.

Risks in New Datacenters as Compared to Telecom Central Offices following NEBS

Everyone agreed that NEBS has been a great success, and that facilities following NEBS are safe with respect to fire risk. Most felt that there was a need for a replacement for NEBS for the new applications (i.e., data centers), and that currently the standards being followed are not as good, and certainly not as uniformly followed as were NEBS. Respondents felt that movement towards a new standard was a good thing. Several felt that the success of NEBS has created some complacency: that the low level of fires is due to the success of NEBS, but with changes due to rapid innovation in Information Technology systems, things are not as safe as NEBS, but there needs to be the same level of commitment to stringent standard to insure continued success. One respondent felt that there are good standards for data processing centers that can be followed, and that some owners are following them.

Risks from Contracted Work

Several of the respondents felt that the more common use of sub-contractors to do work in DPC and TCO sites leads to more variability and greater risk of mistakes and accidents. They felt that the contractors often us lower standards for their training, procedures, workmanship, and materials, and that these are not as tightly controlled as had been the case with the old Bell system. One respondent disagreed, and felt that some of the larger data processing centers are very careful with regard to fire safety procedures in their data processing centers.

Approaches for Specifying a Test Method

Most of the respondents felt that the problem had to be broken down into different categories of fire threats, based on energy input. A few felt that it was still difficult to make it tractable (because of the wide range of conditions), and so picking a few specific examples (or even just one example) to start with, and studying that, would be the best way to move forward. Others contended that the problem is still too widely defined, so it's best to just design a test for which the externally input energy to the burning material is an independent variable, find the sensitivity of the suppression process for a given material to the energy input, and then let the system designers (or Fire Protection Engineers) decide on what electrical systems they can protect with what amounts of suppressant.

Clean Agent Effectiveness in High-Energy Electrically-Energized Fires

Four respondents felt that energized high-energy cable fires should not be suppressed with clean agents, and one more thought the same for "large enough fires."

2.1.3 Case Studies Supplied by FM Global

FM Global graciously supplied three detailed case studies of fire incidents from their experiences, and these are supplied in Appendix II. In the first case study (2006), workers from a sub-contractor were installing a sixth static switch (adding to the five already present). As they were pulling cables under a raised floor, they heard a series of loud noises (described as "three explosions in sequence") coming from one of the five existing static switches in the data processing room. Apparently, wiring inside one of the five existing static power switches overheated and caught fire in an electrical cabinet, setting off smoke alarms. The automatic Halon 1301 system had been turned to manual operation mode prior to the start of work (to prevent a false-alarm release). The system design was for automatic emergency power off in the event of halon release, which did not happen here. Heavy smoke was developing from within the switch's metal enclosure, so employees proceeded to the Medium Voltage room below, and manually tripped the breaker feeding power to the affected room. Employees then opened the cabinet and manually discharged CO_2 extinguishers into the cabinet. Later, the public fire department arrived and fully extinguished the fire, and started ventilating the room. Upon inspection, the fire's thermal damage was found to have occurred in a 10 cm length of group of plastic-insulated cables inside the metal cabinet's enclosure. The source or cause of ignition was not determined.

The second case history (1997) involved an electrical equipment cabinet (3 m long, 3 m high, and 1 m deep) with three bays. The central bay has AC-DC power conditioning equipment, with 208 V 3φ power input, and 12 V, 290 A, 5 V, 500 A output. End bays are data storage bays, and power is delivered around the perimeter of each bay on four buses. Each of the end bays contains eight rows of 4 disk drives per row. Each hard drive is connected to a mid-plane, extending the width of the bay. The hard drives are encased in plastic, and the data storage bays contain polycarbonate (PC), fire retarded PC, and PVC. The three bays are separated by metal sheet, and each bay is ventilated by exhaust fans in the top. The room smoke alarm activated, and the fire department arrived but could not find the fire because of thick black smoke. At about 22 min elapsed time, the emergency power off switch was activated, and at 24 min, the fire department found and extinguished the fire using several 4 kg portable Halon 1211 extinguishers. Inspection revealed that an area 15 cm x 15 cm on the bay mid-plane (presumably plastic) was consumed, and the plastic casing on ten of the hard drives was partially or totally consumed. Copper wiring was found intact, and no melted copper wiring was found. The failure leading to ignition was not reported.

The third case history (1993) involved an automatic voltage regulator in a data processing center. The involved area, the 170 m^2 (1800 ft^2) VAX room, contained four VAX 8000 series computers, and thirty-five RA series disk drive units, other CPUs and modem units. A fire alarm activated three of the four present automatic Halon 1301 systems, and twelve 64.5 kg (142 lb) halon cylinders were released, one in the VAX room ambient, one in the VAX room under-floor, and one in the larger electronic data processing center space, 400 m^2 (4300 ft^2) surrounding the VAX room. Firemen saw that the halon system had activated and saw no flames, so they entered the VAX room. They noted that the fire had occurred in the automatic voltage regulator on a perimetral wall of the VAX room, that the room had been completely electrically de-energized (part of the system design), and that the halon had completely extinguished the fire. According to the FM Global investigators, the voltage regulator (45 kV, 380 V) could have caught fire because of an overheating automatic regulation rheostat. The fire could have propagated from the voltage regulator, through the under-floor cables to the other VAX equipment, as evidenced by the partially burned cables which fortunately were extinguished by the halon.

Table 1 summarizes the FM Global case studies. Voltage supplied to the involved equipment varied from 5 V to 45 kV. In all of these cases, the power was shut down before release of the agent. The agent (CO$_2$, Halon 1211, or Halon 1301), successfully extinguished the fire. In the third case, the specific component which failed is clearly implicated, in others, the cause is unknown.

Table 1 – Summary of FM Global case studies.

Case	Voltage	Material burned	Power-down prior to agent release	Agent (Manual/Auto Release)	Fire extinguished
1	380 V AC	10 cm of plastic cable insulation "10 cm of grouped low-current, plastic-insulated cables inside the metal cabinet's enclosure"	Yes	CO$_2$ (Manual)	Yes
2	12 V/ 5 V DC	10 plastic hard drive cases, 225 cm^2 area of PC, FR-PC, or PVC in the cabinet bay (unclear which material).	Yes	Halon 1211 (Manual)	Yes
3	45 kV/380 V AC	Voltage regulator, cables.	Yes	Halon 1301 (Auto)	Yes

One can't extract any information about the effect of electrically energized equipment in the fires (there wasn't any), but we can look at the effects of flame interaction. In both Case Study 1 and Case Study 2, it does seem that there was interaction between multiple burning surfaces. Hence, in these examples the suggestion of Respondent 18 is validated: that in electrical fires, the initial area of involvement can be bigger initially, so one must consider the classic arguments about including radiant augmentation when assessing material flammability (or suppression of flames on materials).

These case studies provide probably about as much detail as one might hope to get about a fire incident, unless one is involved in an actual forensic investigation, or has unique access to the documents or people involved. Yet it is difficult to use any of the information here to come up with a test method which considers the effect of keeping the system electrically energized while suppressing the fire. For all of these fires, the electrical service was shut off before suppression. In two of the three, there are no details of the failure mechanism itself, let alone estimation of the power levels involved in the failure and the duration of their involvement (which is the information need to design a test procedure which includes the effects of energy-augmented combustion). This is not a criticism of the case studies; we are very fortunate and indebted to FM Global for providing these materials. Rather, to provide the level of detail of information which we need for our task, the investigators at the site probably have to go into their investigation intending to extract information specifically about the topics listed in Table 2.

Table 2 – Useful questions for forensic fire investigators to keep in mind to when gathering information useful for understanding suppression of electrically energized equipment fires.

1. What power level was involved?
2. For what time period was it involved?
3. Was there any electrical involvement just an ignition source, or did it add energy to the burning material, before, during or after the ignition?
4. Did the energy-augmentation continue during suppression?
5. What was the configuration of the electrical energy release?

Given the difficulty in even determining the source of the fire, one would be very lucky to get this detailed information from an incident report. Nonetheless, with material and configuration data, it might be possible to estimate the amount of heat feedback from the electrical source to the area of burning material.

One approach to getting more detailed information in future studies would be to identify someone who has done a lot of forensic studies of electrically-induced ignitions in data processing or telecommunications equipment. If asked to keep in mind the questions of Table 2, they would probably be a good source of information in the future.

2.2 Literature Review

The areas of material flammability and flame suppression are too big to review here, but some background is provided in areas which are important for the suppression of electrically energized equipment fires.

2.2.1 Materials Flammability

The burning of solid materials in a fire is a complex and well-studied phenomena, yet simple descriptions are available in the literature [16-18]. A description based on heat balance at the surface is illustrated in Figure 1. Heat input comes from convection and radiation from the hot flame over the polymer, as well as from any external source. These external sources include radiation from adjacent flames, hot upper layers, arc discharges, or radiant heaters, as well as conduction from hot surfaces (e.g., overheated wires, resistive heaters) or from adjacent hot gases. Heat losses from the system include reflection of incoming radiation, re-radiation of the hot polymer surface, and conductive losses into the interior of the polymer. The heat conducted into the interior of the polymer is a loss (for short times) since the energy may not yet be contributing substantially to the mass loss; at larger times, that energy comes back out, as the material burning at later times is essentially preheated.

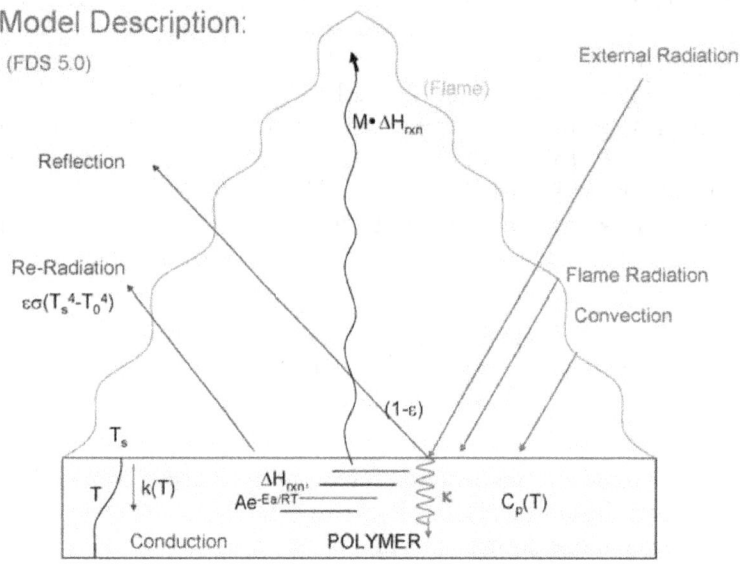

Figure 1 – Heat flows at the surface of a burning thermoplastic polymer, with some of the processes in the condensed phase illustrated also.

If the heat gains and losses are summed into a net heat input to the polymer q_{net}", the mass loss rate at the surface can be described by:

$$\dot{m}" = \frac{q_{net}"}{L_v},$$

Eq. 1

in which $\dot{m}"$ is the mass loss rate per unit area, $q_{net}"$ is the heat input per unit area, and L_v is the latent heat of phase change/decomposition. The net heat input $q_{net}"$ is defined by:

$$q_{net}" = q_{f,rad}" + q_{f,conv}" + q_{external}" - q_{re-rad}" - q_{poly,conv}" - q_{pol,cond}",$$

in which $q_{f,rad}"$ and $q_{f,conv}"$ are the radiation and convection heat transfer from the flame to the polymer, and $q_{external}"$ is the externally applied radiation. The re-radiation heat losses from the polymer to the ambient is given by $q_{re-rad}"$ which is equal to $\varepsilon\sigma(T_{pol,surf}^4 - T_{amb}^4)$; $q_{poly,conv}"$ is the convective heat losses from the polymer surface to the ambient, and $q_{pol,cond}"$ is the heat loss into the polymer by conduction.

An illustration of the effects of different heat input rates on the burning of a polymer [19] is shown in Figure 2, where a 25.4 mm thick slab (1-D) of poly-methylmethacrylate (PMMA) is subjected to external heat fluxes of 10 kW/m^2 to 70 kW/m^2; Figure 3 shows the same data for earlier times. As Figure 2 shows, samples subjected to higher fluxes have a higher average mass loss rate, and a shorter burning time. The shape of the curves is also different. At 20 kW/m^2, the mass loss rate barely reaches a steady-state, and at 70 kW/m^2, the peak mass loss rate at the end of the burning period is very high. These effects are caused by conduction into the polymer. The transient in the beginning is caused by conductive *losses* into the polymer, while the peak at the end results from heat *gains* as the heat previously conducted into the polymer has raised its temperature (effectively preheating the polymer), so that it has a higher burning rate.

The differences in the mass loss rates at early times as shown in Figure 3 result from two causes. The conductive heat losses (around 4 kW/m^2) are a bigger fraction of the total heat input for the low flux cases, so the energy left to cause mass loss is much smaller. Also, the mass loss itself causes regression of the polymer surface, which affects development of the temperature profile. That is, there is a thermal wave propagating into the polymer, as well as a surface regression rate, which are inter-related. The temperature profiles as a function of time are shown in Figure 4 and Figure 5, for 20 kW/m^2 and 70 kW/m^2 external heat input. In Figure 5, the cluster of overlapping temperature profiles near t=400 s corresponds to the steady burning period (200 s to 500 s in the 70kW/m^2 curve of Figure 2). As Figure 4 and Figure 5 show, the temperature profiles reach a steady state much faster in the high flux case. This is not because the energy is conducted in faster (the surface temperature is about the same in both cases and the thermal diffusivity is about the same); but rather, the surface is swept away more rapidly in the high-flux case, allowing a steady-state temperature profile to develop, which is present until the thermal wave reaches the back side of the sample and the entire remainder of the sample heats to the decomposition temperature.

The significance of these results for the case of energy-augmented combustion is two-fold. First, at low flux (i.e., at early stages of burning when the heat feedback from the flame is small), the mass loss rate is very sensitive to any additional heat input since the conduction losses (as well as re-radiation losses) are a large fraction of the net heat (which may not even be greater than zero). Hence, additional heat from a radiative source or an electrical short will have a big effect. Second, if suppression tests are performed on a solid sample, the net energy flow into the polymer is affected by the heat losses, and these in turn are influenced by: preheating from the flame, preheating from any external energy source, thickness of the sample, and time for initiation of the suppressant flow. Hence, these influences must be carefully considered in the test procedure. Of course, these effects are magnified geometrically since a burning solid sample is a positive feedback system: heat feedback increases the mass flow of fuel, which makes the flame bigger, which increases the heat feedback.

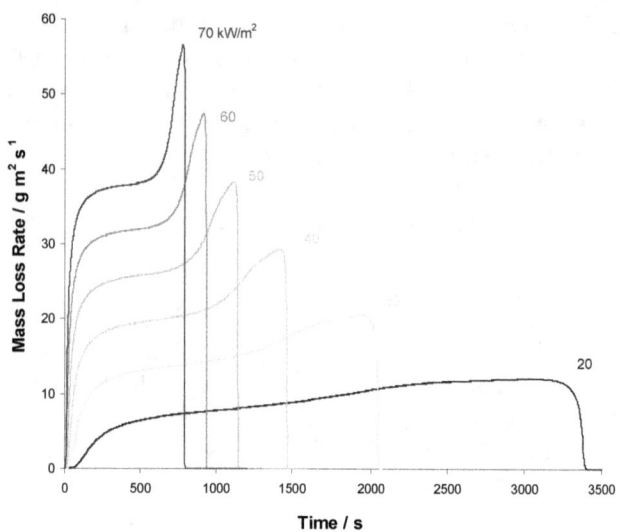

Figure 2 – Calculated mass loss rate of 25.4 mm thick PMMA as a function of time for incident external flux rates of (10, 20, 30, 40, 50, 60, and 70) kW/m².

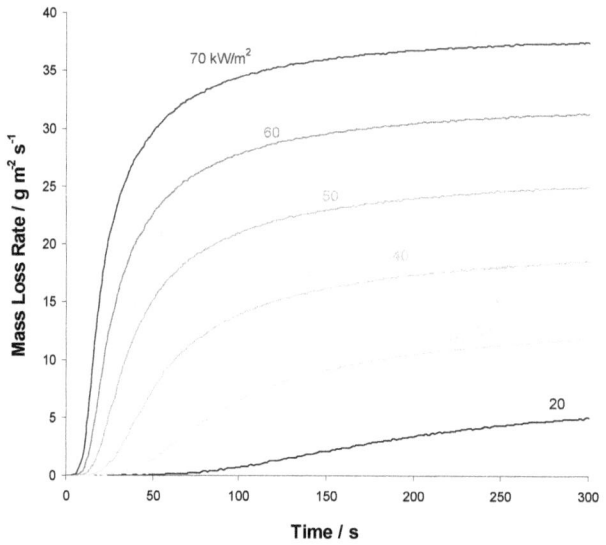

Figure 3 – Same data in Figure 2 but at shorter times.

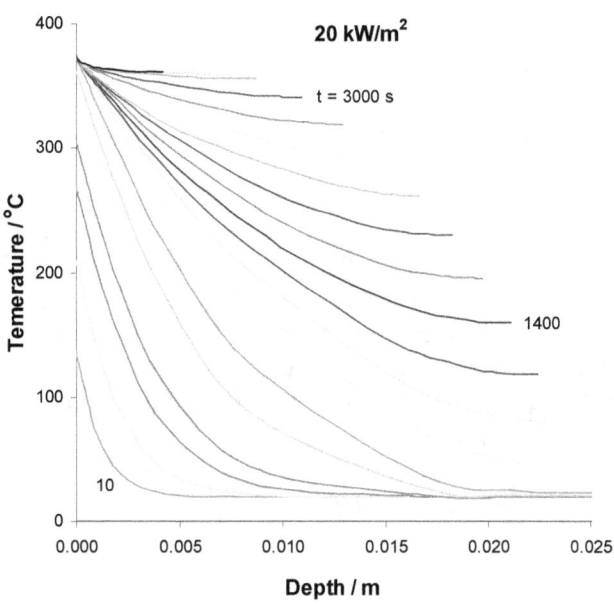

Figure 4 – 1-D Temperature profile through PMMA slab (initially 25.4 mm thick) as a function of time (indicated on curves), at incident flux or 20 kW/m^2.

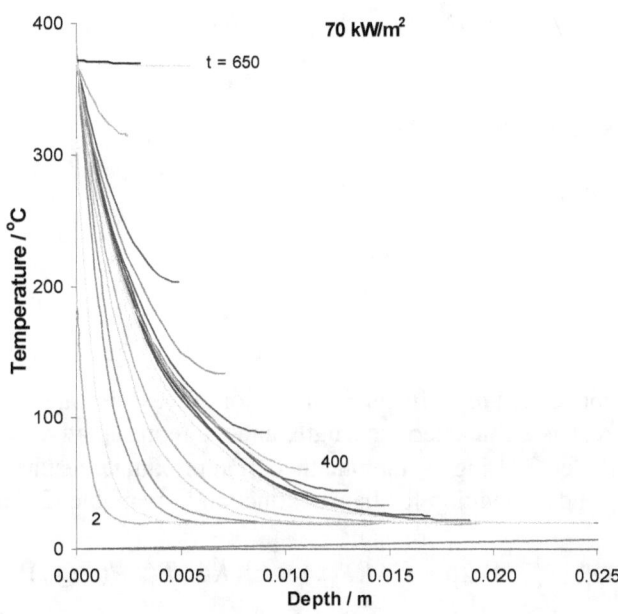

Figure 5 – 1-D Temperature profile through PMMA slab (initially 25.4 mm thick) as a function of time (indicated on curves), at incident flux or 70 kW/m^2.

2.2.2 Fire Suppression

Unified View of Fire Suppression

Simple models of flammability (and hence suppression of fires) were based on the fire triangle: fuel, oxidizer, and heat are needed to maintain a flame [20]. Upon development of the brominated fire suppressants, this was extended to be the fire tetrahedron, in which chain reaction (i.e., robust concentrations of chain-branching radicals) was also a requirement for fire. A more comprehensive description of fire suppression was described by Williams [21], in terms of the characteristic chemical reaction time, τ_c, and transport (i.e., flow or diffusion) time, τ_r. In general, the chemistry must be fast enough to keep up with the flow field effect, or the flame will extinguish. This process is described in terms of the Damköhler number, $D \equiv \tau_r / \tau_c$, which is the ratio of the characteristic flow residence time to chemical reaction time, or alternatively, the ratio of the chemical rate to the transport rate. The chemical rate is given by an Arrenhius-type expression:

$$w = c_F^n c_O^m A^{-1} \exp(-E/RT),$$

Eq. 2

in which w is the reaction rate, c_F and c_O are the concentrations of fuel and oxidizer, A and E are the Arrenhius collisional term and activation energy, and T is the temperature. The chemical reaction time τ_c is the density divided by the volumetric reaction rate

$$\tau_c \equiv \rho/w = \rho \, c_F^{-n} c_O^{-m} A \exp(E/RT).$$

Eq. 3

The characteristic flow residence time is either

$$\tau_r = \ell / v$$

Eq. 4

or

$$\tau_r = \ell^2 / \mathcal{D},$$

Eq. 5

depending upon whether convection or diffusion is the major process of transport into the reaction zone during the extinction. Here, ℓ is a characteristic length, and v a representative velocity, and $\mathcal{D}$ is an appropriate diffusion coefficient. Using asymptotic theory, approximate results with general applicability have been developed [22], and a condition for flame extinction is available [21] as:

$$(l^2 / \rho\mathcal{D}) \, c_{Fb}^n \, c_{Ob}^m \, A \exp(-E/RT_{AF}) < k \, [(RC_p T_{AF}^2)/(EQ_F)]^3$$

Eq. 6

where T_{AF} is (approximately) the adiabatic flame temperature, k is a constant (usually around 10^{-3}), C_p is the average specific heat at constant pressure for the gas phase, Q_F is the heat released per unit volume in the gas phase, and b denotes conditions at the system boundary (i.e., inlet). The significance of this framework is that all of the approaches for fire extinguishment:

1. cooling the gas phase,
2. cooling the solid phase,
3. isolating the fuel.

4. isolating the oxidizer,
5. inhibiting the chemical reactions, or
6. blowing away the flame

can all be described analytically by the above equation. Anything which lowers the left side of Eq. 6 enhances extinction, for example, reducing the temperature (lowering T_{AF}), lowering the concentration of fuel C_F or oxidizer C_O, or cooling the condensed phase (also lowers C_F). The form of Eq. 6 relevant for convective flow control (rather than diffusion) replaces $\ell^2/\mathcal{D}$ by l/v, so that increasing the convective flow (i.e., blowing on the stabilization region), increases v, and again lowers the left-hand side of Eq. 6 and enhances extinction.

Flow-Field Effects

As an illustration of these effects, the results for Halon 1301 and halon replacements added to the air stream over opposed-flow heptane–air diffusion flames is shown below [23]. In the experiment, the oxidizer is directed down (stagnation flow) against a 50 mm diameter pool of heptane. The oxidizer flow velocity is set, and agent is added to the air stream until extinction occurs. If the velocity of the oxidizer flow is increased (i.e., the flow residence time decreases), the amount of agent required for extinction also decreases. Figure 6 shows the extinction mass fraction of the suppressant in air as a function of strain rate, a, for gas inlet temperature of 25 °C. The strain rate (s^{-1}) is the normalized velocity gradient along the streamline dv/dy; where $v = -ay$ for a stagnation flow, so a is proportional to v. Curves are shown for a large number of agents. As indicated, higher gas velocities (strain rates) require a lower agent mass fraction for extinction. Figure 7 shows the comparable data for gas inlet temperature of 150 °C. As indicated, higher gas temperatures require more agent (at all flow velocities). This is because, as described by Eq. 6, as the inlet temperature goes up and the left hand side of the equation goes up, making the flame harder to extinguish.

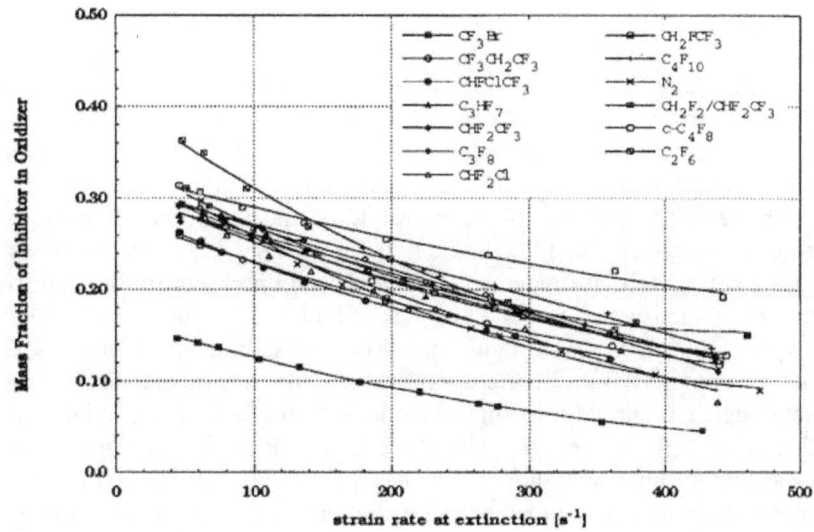

Figure 6 – Mass fraction of inhibitor in the oxidizer flow (air) necessary to extinguish a counterflow diffusion flame over heptane, as a function of strain rate ($T_{air, inlet}$ = 25 °C).

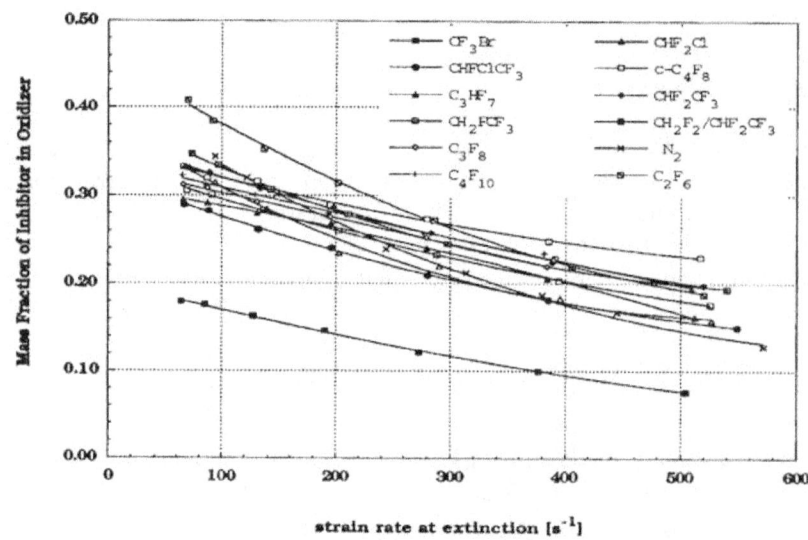

Figure 7 – Mass fraction of inhibitor in the oxidizer flow (air) necessary to extinguish a counterflow diffusion flame over heptane, as a function of strain rate (T$_{air, inlet}$ = 150 °C).

One might ask what amount of agent would be required if there were no limiting characteristic flow time (i.e., at zero strain rate). This would represent a condition for which all flames of the given mixture would be extinguished. One can arrive at this value by extrapolating the above curves to zero strain rate at extinction. Alternatively, these have been obtained in premixed systems as the inerting concentration of agent for all values of the fuel/air mixture (stoichiometry) for a particular fuel [24]. As the inlet temperature of the mixture increases, the flammability limits widen, and the amount of agent required for inertion increases.

For many flames, a suppressant is added at concentrations much lower than the inerting concentrations, and the flame extinguishes because of local flame destabilization and blow off. That is, there is a crucial location in the flow-field, the stabilization point, where addition of a suppressant causes the characteristic chemical time to become larger than the characteristic flow time; the chemistry can't keep up with the flow, and the flame extinguishes at the point (blows off). An example of this is the cup burner flame, for which the blow-off extinguishment has been found to occur due to destabilization of the flame at the base region [25]. As can be seen in Figure 8 below (from ref. [26]), the cup burner flame <u>blow-off</u> concentrations (lower set of symbols and line in the figure, for CO$_2$, N$_2$, Ar, and He) are significantly lower than the <u>inertion</u> concentrations (upper set of symbols and line). For flames in microgravity, however, where the strain rate is very low (due to a lack of buoyancy-induced flow), the flames are much more robust and require more agent for extinguishment (i.e., the stabilization is not upset by the buoyancy-induced flow near the base). In microgravity, the flame tip extinguishes before the flame blows off, and the amount of agent required for extinguishment in microgravity is essentially equal to the premixed flame flammability limits measured elsewhere [24]. That is, in microgravity, without flame base oscillation caused by the buoyancy-induced vortices [27], the flame stabilization is much better, the flame requires about 43 % more agent for suppression, and the suppression concentration is essentially the inerting concentration. Hence, for a particular flame configuration, it is very important to consider the flow field and flame stabilization when considering the apparent extinguishing concentration.

18

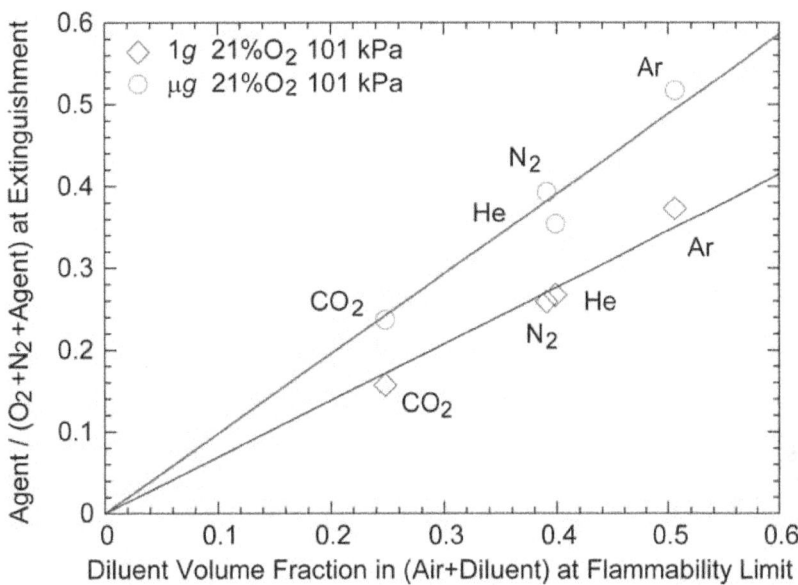

Figure 8 - **Correlations between the cup burner extinguishment limits and the critical flammability limits expressed in terms of the agent volume fractions in oxidizer, from [26].**

Effects of Heat Addition on Suppression

Within the framework described above, the effects of energy-augmented combustion on clean-agent suppression of flames over condensed-phase fuels can be understood clearly. Adding energy to the condensed phase increases C_F, and adding it to the gas phase increases T_{AF}, in these cases reducing the likelihood of extinction. Adding a halogenated clean agent which decomposes in the flame (CF_3Br or CF_3H) [28], lowers the overall reaction rate [29,30] (lowers A or raises E), lowering the left-hand side of Eq. 6 and enhancing extinction. Increasing the flow velocity (i.e., blowing on the stabilization region) decreases the flow residence time (decreasing the left-hand side), again enhancing extinction.

In both experiments and detailed numerical modeling, a higher temperature flame requires more agent for extinguishment, with either chemically reacting or inert agents. For example, extinguishing heptane cup burners by addition of Halon 1301 can require 2.5 times as much agent when the oxygen volume fraction in the air goes from 0.21 to 0.286 [31], and CF_3H can require 1.75 times as much when the oxygen volume fraction goes from 0.21 to 0.264 [32]. Likewise, for extinguishing methane-air cup burner flames with CO_2, about 2.1 times as much agent is required when the oxygen volume fraction goes from 0.21 to 0.30 [33]. For the inert agents with higher O_2 volume fraction, the larger agent volume fraction is required to reduce the flame temperature to the same equivalent value at which extinguishment occurs [33]. For the chemically-active agents at higher oxygen volume fraction, the flame temperature is higher, causing higher radical concentrations, which then require more agent to bring them down to the levels at which extinguishment occurs. Figure 9 shows the variation in the final flame temperature and volume fraction of H atom due to changes in the oxygen content of the oxidizer stream, for a premixed $CH_4/O_2/N_2$ flame. For oxygen volume fractions increasing from 0.2 to 0.3, the adiabatic flame temperature rises 354 K, from to 2181 K to 2535 K, while the final [H] goes from 260 μL/L to 3200 μL/L. The significance of these findings, to the present problem of energy-augmented combustion in electrically-energized equipment, is that for premixed or diffusion flames, higher gas-phase temperatures will require higher agent concentrations for extinguishment. Alternatively, if the energy is added to the condensed phase, a larger flame results, and the heat losses represent a smaller fraction of the total, so that the gas-phase temperature will again rise.

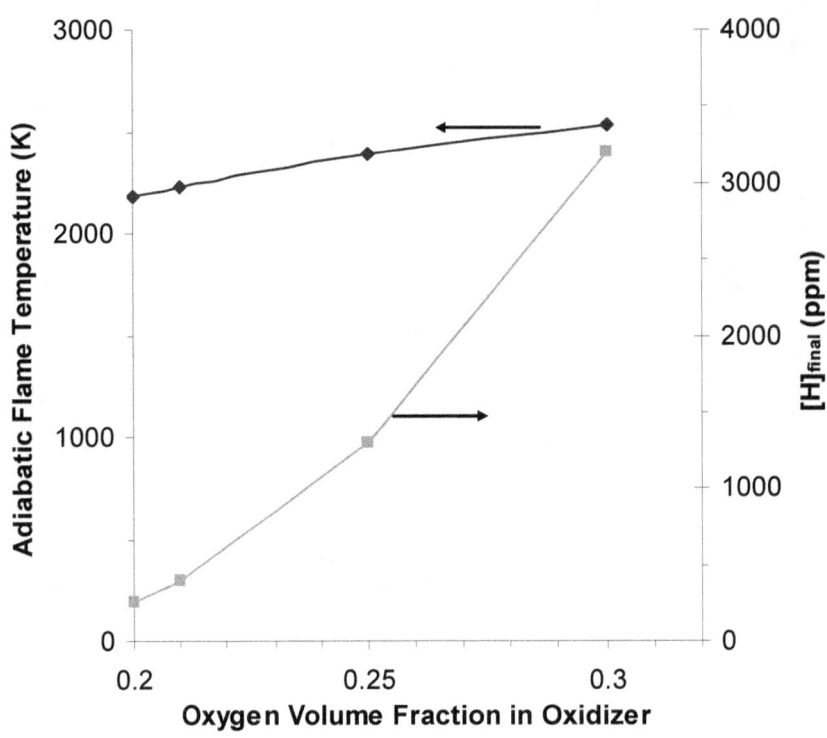

Figure 9 – Adiabatic flame temperature and final H-atom volume fraction as a function of oxygen volume fraction in the oxidizer in a CH₄/O₂/N₂ premixed flame.

2.2.3 Suppression of Flames Over Condensed-Phase Materials:

Existing standard (or nearly standard) tests for fire suppression include the Limiting Oxygen Index (LOI), cup burner, NFPA 2001, the UL tests referenced in NFPA 2001, and the pan tests for fire extinguishers. In the LOI test [34-36], a polymer sample is held vertically in an oxidizer flow, and the oxygen volume fraction necessary to just maintain combustion is noted. This is equivalent to a suppression test, since the volume fraction of added nitrogen (to air) needed to extinguish the flame is calculated directly from the LOI (oxygen volume fraction in the oxidizer) as $X_{N_2} = 1 - 4.76$ (LOI). In the cup burner test [1,37,38], the fuel is a liquid in a fuel cup, or a gas issuing from straightening screens in the cup (31 mm outer diameter), located concentrically in an 85 mm inner diameter chimney, through which air and agent flow at a specified mixture and velocity. The minimum extinguishing concentration (MEC) of suppressant required to extinguish (i.e., blow off [39]) the flame is determined. As mentioned above, the NFPA 2001 flame extinguishing concentration for Class B fires is specified by the cup burner MEC values times a safety factor of 1.3. For Class A fires, UL 2127 (inert gases) and UL 2166 (clean agents) are referenced, and the numbers from those tests are multiplied by a safety factor or 1.2. In the UL tests, a large (100 m³) enclosure is used, with the fuel centered in the enclosure and located approximately 20 cm from the floor. Fuel arrays consist of either four vertical plastic sheets (20.3 cm x 40.6 cm x 0.953 cm; PMMA, PP, and ABS) spaced 1.27 cm to 3.18 cm apart; or a wood crib, composed of four layers of six kiln-dried spruce or fir blocks (3.8 cm x 3.8 cm x 46 cm). Ignition is by a pan of heptane burning for six minutes (for the wood crib) or 90 s (for the polymer sheets). In the pan tests for fire extinguishers (ASNI/UL 711), large heptane pan fires of various sizes (0.2 m² to 4.65 m² for indoor tests), or (7 m² to 18.6 m² for outdoor tests), or large wood cribs (72 to 400 block, increasing sizes), are extinguished manually. There are no agreed-upon tests for Class C fires, but a number of tests have been proposed, as discussed below in section 2.4.2.

Several features of the suppression of flames over condensed-phase materials are noteworthy in comparison to the similar suppression of flames over gaseous fuels. First, necessary heat lost to the surface (to provide the fuel) weakens the flame (due to a lower flame temperature). With agent addition, the flame lifts off (to allow more mixing time, better premixing, and a stronger flame [27]), but also leads to less fuel supply—which further weakens the flame, and makes the heat losses more important. This is illustrated in Figure 10 and Figure 11, which show the cup burner heptane or methanol consumption rate as a function of CO_2, CF_3Br, or R-125 volume fraction [40,41]. As indicated, the fuel consumption rate drops very rapidly as the agent concentration nears the extinguishing value. Although agent addition affects the flame temperature somewhat (at near-extinguishment concentrations CO_2 lowers the peak flame temperature of methane-air flames by about 200 K, and CF_3Br raises it by 30 K [42]), the main cause of the reduced liquid fuel consumption rate is likely to be flame stand-off. As indicated in Figure 12, for methane-air cup burner flames, the flame base distance from the burner is seven (or three) times higher with CF_3Br [42] (or CO_2 [27]) added at near-extinguishment concentrations. This lower heat release rate near extinguishment has also been shown in flames over condensed-phase materials in a cup burner like configuration [43].

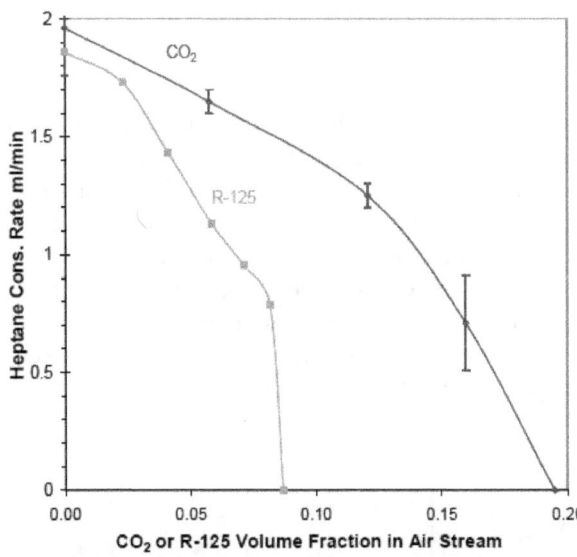

Figure 10 – Cup burner heptane consumption rate as a function of CO_2 or R-125 volume fraction in air.

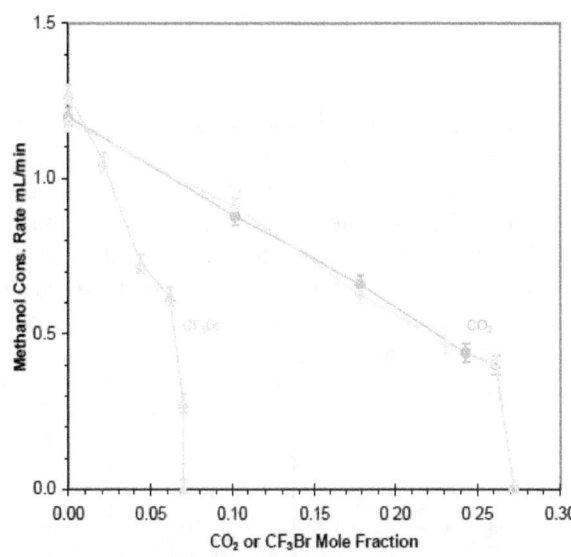

Figure 11 - Cup burner methanol consumption rate as a function of CO_2 and CF_3Br volume fraction in air.

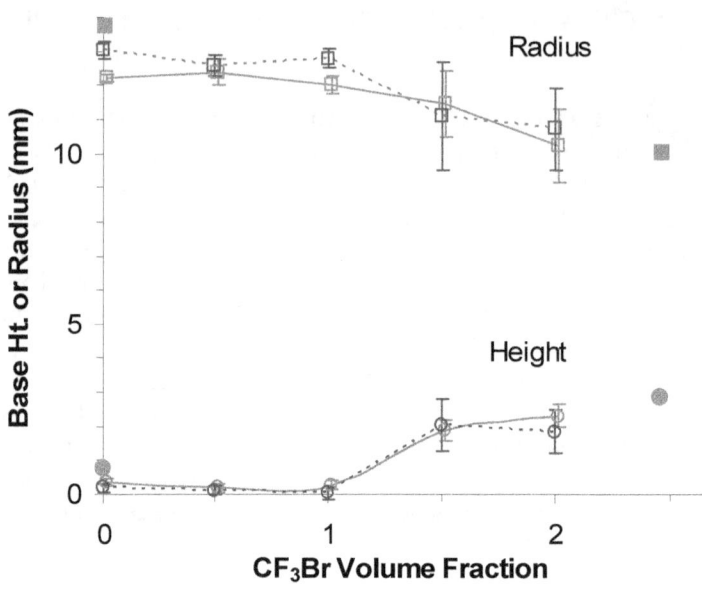

Figure 12 – Flame height and radius for methane-air cup–burner flames with CF₃Br added to the air stream at increasing volume fraction [42].

Another difference in suppression of flames over condensed-phase materials is that the flame stabilization process is intimately connected to the material configuration. Since the flame must exist near the surface of the burning material (to supply the heat feedback necessary to supply the fuel), the configuration of the material affects the flame stabilization, and hence, the amount of agent necessary to extinguish the flame. For example, in recent experiments, Takahashi et al. studied cylindrical burners with methane issuing from porous surfaces [43]. The cylindrical porous burner was oriented for either radial fuel supply (from a continuous rod), end up, or end down, as shown in the left, middle, and right images of Figure 13. The first configuration, radial fuel supply, was the most stable, and required 23 % more CO_2 for suppression that the end-up fuel supply (which were very close to the methane-air cup burner values).

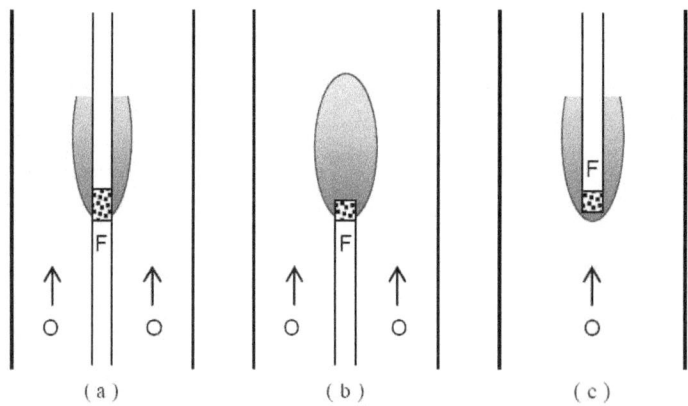

Figure 13 – Cylindrical fuel configurations methane-air porous burners (from ref. [43]; used with permission of the author).

Other effects are also important for the suppression of flames over condensed-phase fuels. Because the fuel supply rate is dependent upon material temperature (which drives the decomposition), preheating of the material has a large effect on the fuel supply rate, and the amount of agent necessary for suppression. This has been shown by Goldmeer and Urban [44] and Ruff et al [45] for flames over cylindrical PMMA. Also, melting and dripping also occur for solid fuels, and these effects can both change the shape of the burning surface (affecting stabilization and heat transfer from the flame and any auxiliary source), as well as draw energy from the reaction zone [43,46,47].

Because of the intricate balance of heat flows as described in section 2.2.1 above, any changes in the net heat flux near the critical values for ignition have a large effect on the burning behavior. This is illustrated in Figure 14, which is a flammability diagram from PMMA [48]. The figure shows the following, as a function of the imposed heat flux from a radiant heater: the flame spread rate, upward or downward (left two curves); the steady mass loss rate (upper right curve); and the time to ignition (lower right curve). As indicated, changes in the net heat flux near 8 kW/m^2 have a huge effect on the flame spread rate and the time to ignition; of course, these results will vary with polymer type. It would be immensely valuable in the context of the current work to have such diagrams in the presence of increasing amounts of gas-phase suppressants for materials used in electrically energized equipment.

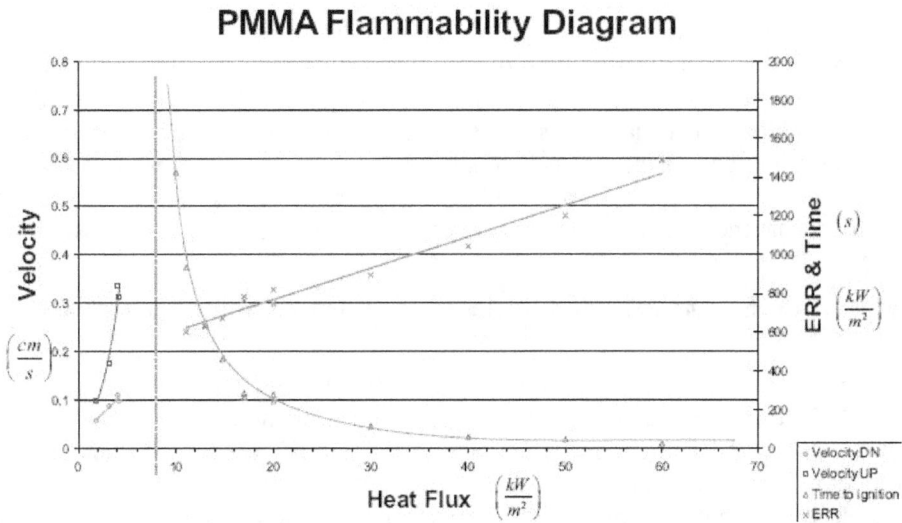

Figure 14 – Flammability diagram for PMMA; (from ref. [48]; used with permission of the author).

2.3 Threat Definition

In order to define critical applications, equipment types, fire threats, potential clean agent applications, agent discharges, and reported incidences, surveys are planned for the future [49]. At present, we have the results of the discussions with technical experts, and the case studies supplied by FM Global (as outlined above and described in the Appendices). These provide useful insights into the likelihood of obtaining enough detail in a survey to understand the power input to, and its influence on, a fire in electrical equipment or cables.

In the interviews with technical experts, the detail in the information provided varied widely, and this is likely to occur in a survey as well. The relevance and amount of information varied partly because of the recent experience of the interviewees (for example, while all are undoubtedly experts in the general area, some related that they have moved away from the detailed knowledge of the systems and applications which they had in the past). In other cases, it may not be appropriate for them in their current position to release potentially proprietary information without a large, formal review process.

The findings of the present study are relevant to any survey which may be conducted in the future. First, it is important to understand at the outset what information a survey is likely to provide. A survey can give some statistical data on suppression releases, suppressants used, fire incidents, equipment involved in the fire, success of suppression releases, and electrical shutdown in the event of a fire. In general, databases on fire incidents in electrical equipment have few data points and few details, such that often, even the cause of ignition is not available in a fire incident database. In a new survey, it is possible that individuals with enough detailed knowledge of the failure mechanism could be accessed; however, whether they could share their knowledge is unclear. If the survey merely accesses existing databases, it is unlikely, a priori, that the database will have information on the exact failure mechanism (and the likely power dissipated during the event), necessary to assess the role of energy-augmented combustion. Hence, for the purpose of defining the threat, surveys are useful but not sufficient.

Based on the literature review, we have defined the physical phenomena in the burning material which can influence the quantity of agent necessary to extinguish a fire. Table 3 lists the questions which must be answered in order to identify which of the relevant parameters are present in the fire to be suppressed, and hence, to develop at test method to accurately represent the threat in the field. This list can serve as a guide for either a survey which seeks to identify what fire threats have occurred in the field, or laboratory tests or forensic investigations which can be conducted to understand the fire incidents which have occurred. Of course, since the actual values of the parameters may vary, one approach could be to adopt more conservative values and let the test method apply to a wider range of scenarios; another approach could be to estimate the actual value of each controlling parameter for each situation (i.e., a performance-based approach), so that the fires protection resources might be employed most efficiently.

Table 3 – Questions which are necessary to answer about a fire in electrical equipment useful in developing a relevant test method.

1. Is there energy addition from an external (i.e., electrical) source or not?
    a. If there is energy addition, how much, and for what duration?
    b. Is there pre-heating of the material prior to ignition? If so, how much?
2. How does the ignition occur?
    a. Is autoignition required, or is there a separate ignition source?
    b. If separate ignition source, what is its assumed duration?
    c. What are the characteristics of the initial ignition event with regard to size of initial flame?
3. What is the material burning?
4. What is the configuration of the burning material
    a. adjacent materials,
    b. orientation,
    c. temperatures,
    d. confinement of melting and dripping materials.
5. Is there involvement (i.e., heat feedback) from adjacent flames?
6. What is the ventilation condition in the burning area?
    a. velocities,
    b. stabilization condition of the flames?

As for threat definition based on the survey of technical experts, there was general agreement that the possible scenarios for energy-augmented combustion are very wide, and could not be addressed together.

Many of the experts suggested that perhaps the failure scenarios could be categorized according to the possible power level in the equipment, or by the many types of failure. One respondent suggested that even that approach was too ill-defined, and that to be tractable, a good approach would be to consider only one failure mechanism for one type of equipment, understand that, and then move on to another case, then another, then another. There was a broad agreement that adding energy to the burning system would cause it to require more suppressant for extinguishment. Several respondents thought it was clear that there are some situations in which energy is clearly added (e.g., arcing cable fires), and that the current design concentrations probably would not work for that situation.

All of the fires related in the phone interviews (and the FM case studies) are listed in Table 4. Most of the fires described occurred in power equipment (batteries, cables, power switches, etc.). In none of the fires was suppression attempted with electrical power still on. With more time, the energy fluxes in all of the fires reported by Respondent 05 might have been discernable (and they were mostly huge, intense events). In none of the other reported events was there enough information available to quantify the energy flux to the burning material. These conversations are probably at the higher end of information typically available. As described later, threat definition will probably require access to individuals performing forensic analysis, or laboratory re-creations and modeling of failure events based on statistical data and expert input.

Table 4 – Summary of Fire Events Described by Respondents.

#	Description	Source	Voltage	Power kW	Material burned	During servicing	De-powered?	Extin. method
Batteries								
1	cracked battery, leaked, shorted, burned	2	DC	high	battery jar	?	Self	SE
2	cracked battery, leaked, shorted, burned	5	540 V DC	400	PP battery jar, conductor failure	Y	Self	SE
3	Attached reverse polarity, shorted	5	48 V DC	high	battery jar, conductor failure	Y	Self	M, H_2O
Rectifiers, varistor, power electronics								
4	Rectifiers, varistor, power electronics	2	?	?	?	?	?	A, 1301
5	Power-shutdown induced failure of varistor	5	? DC	High	Varistor	Y	na	SE
6	Fire in automatic voltage regulator	F3	45 kV,AC 380 V,AC	High	cable insulation	N	Y	A, 1301
Cable Fire								
7	Overheating of a group of plastic insulated low-current wires in a static power switch	F1	380 VAC	?	10 cm of plastic cable jacket	Y	Y, M	M, CO_2
8	Cable connector failure in cable tray	5	DC	High	?	?	?	?
9	Power conductor shorted to fire stop material	5	?	High	Composite fire stop material	?	?	?
AC/DC power Supply								
10	DC Power Plant	5	48 V DC	2400	?	Y	Y,M	
11	Bus-bar accident	5	13.8kV, 600, 347 VAC	4000	Little	Y	na	SE
12	Interlock device failure	5	240 V 3ϕ	400	?	?	?	?
Data Processing Equipment								
13	Fire in hard drive bay	F2	5V, 12V	?	Plastic encl. + hard drive cases	N	Y, M	M, 1211
14	Power supply failure	10	?	?	Smoke, no flames.	N	Y,A	A
15	Power receptacle wiring error	10	120 V	?	Smoke, no flames	N	N	?
Other								
14	Transformer fire	5	high AC	100	6" of windings	?	?	?
15	Return air fan motor (7.5 kW) burnout	5	AC	?	?	?	?	?
16	Overheated Diesel Exhaust (several)	5	na	na	?	?	?	?

2: respondent 02; 5: respondent 05; 10: respondent 10;

F1: FM case study 1; F2: FM case study 2; F3: FM case study 3

A: automatic; M: manual; 1301: Halon 1301; 1211: Halon 1211;

Y: yes; N: no; na: not applicable; Self: self de-powered; SE: self-extinguished;

2.4 Test Method Evaluation and Development

2.4.1 Performance- vs. Prescriptive-Based Approach

To specify a test method, there are two basic approaches. The first is scenario-based, and the other is performance-based. In the scenario based approach, one seeks to characterize individual fire threats, as in the work of Keski-Rahkonen [50]. In that work, they studied ignition phenomena in electrical equipment through statistical data on failures followed by laboratory and modeling studies of the failure mechanism. In the present work, however, rather than studying ignition, the goal would be to understand the role of energy-augmented combustion in the *fire suppression*. Based on statistical data or surveys, one would identify a likely failure mode, and then through laboratory experiments and modeling, study the fire characteristics with regard to the relevant parameters which control the suppressant concentrations. (Table 3 questions). Following these steps, a test method would be designed to reproduce the values of each of the relevant parameters in Table 3, and the suppressant levels could be determined for each agent based on the threat to be protected. Since equipment types and the failure modes are quite different, this approach would have to be applied on a case-by-case basis. After enough understanding was developed, the cases could be grouped, and the suppressant requirements for each group of expected failure types could be specified. Of course, a single worst-case scenario could be identified and used as the test method to specify suppressant requirements for all electrically energized fires (but as with all prescriptive-based codes, this could lead to an inefficient use of suppression resources).

In a performance-based approach, one would specify a test procedure which includes the important parameters that control the suppressant concentrations (Table 2 questions: energy addition, ignition duration, materials, configuration, ventilation, etc.), and then determine the amount of agent needed, based on values of each important parameter. It would then be up to the system designer to (or a Fire Protection Engineer) to determine, for a given application, the value of each parameter for the range of possible failure modes, and hence the amount of clean agent they would need to protect the equipment in the event of failure.

In either case, the goal is to understand first the actual values of the relevant parameters. The only difference is when the values of the relevant parameters are specified: 1.) prior to the *test method* use, or 2.) later when the system is designed and installed. One value of the latter approach is that it will emphasize, to the system designer, that the fire safety can be achieved either through initial design of the equipment, or by post-development suppression of fires. In either approach, the prescriptive or the performance based, it is most useful if the test procedure provides fundamental fire performance data which can then be used to predict the agent performance for a range of conditions. This is desirable in the prescriptive-based approach, since for electrically energized equipment fires, there is such a wide range of values of the possible energy fluxes that could occur, and these will affect the suppressant requirements. Hence, one would not want to have a separate test for each possible failure mode; it would be much more efficient if one test could apply to many scenarios, based on the value of one test parameter (e.g., the imposed external heat flux).

2.4.2 Analysis of Previous Test Methods for Suppression of Energized Electrical Fires
Overview

Previous work to develop a test procedure to simulate the suppression of electrically-energized fires can be grouped into three categories. The first is based on the failure mechanisms deduced from the fire incident reports available in the statistical databases. The advantage of this technique is that from the start, it attempts to use conditions which are representative of the actual fire threats to be extinguished by the clean agents. The tests are essentially attempting to simulate a failure mechanism believed to be representative. The second category is based on creating conditions which control the most important parameter (for example, the external energy added to the burning polymer, or the autoignition temperature in the presence of suppressant), and quantifying the response of the suppression process to changes in that parameter. While little attempt is made a priori to correlate this parameter with its relevant value in suppressed Class C fires in the field, the advantage of this technique is that the most important parameter, the external heat flux, is carefully controlled and quantified in the experiments. For example, the tests are conducted over a range of external heat fluxes, so presumably the results can be used to understand the effects of this external heat flux on a wide range of conditions which may be present in actual suppression of electrically-energized fires. The tests allow better ranking of agents than does a pass/fail test, and since the measurements provide fundamental parameters, the results might eventually be used in performance based design calculations. Finally, a third category is a miscellaneous assembly of other test methods. Many of these can be placed in one of the first two categories, but in some cases they studied ignition (instead of suppression), or they did not add energy per se (although the results are informative for the present discussions). The relevance of these tests is better highlighted by keeping them in a separate category.

The first category, Tests Simulating the Failure Mechanism, includes work in the early 1990s at FM Global to simulate an arc discharge with copper-coated carbon rods with Halon 1301 added to suppress attached flames. Others developed a series of tests which aimed to simulate actual failing components in telecom and data processing equipment. These included work in the late 1990s by McKenna et al. [4,5] which simulated ohmic heating of wire bundles, and conductive heating of an isolated wire, and printed wiring board arcing failure. Soon thereafter, Niemann and co-workers [14,15,51] suggested a modification to the conductive heating test to include a continuous ignition source. Work in the late 2000s by Stilwell and co-workers, simulated an overheated hot wire in the vicinity of a flammable polymer [6], or embedded in a wire bundle [7]. Since most of these approaches are scenario based, they would lead to a standard test method that is prescriptive.

The second category, Test Methods Based on Controlling the External Heat Flux, includes work in the 1970s by Tewarson and co-workers using the Fire Propagation Apparatus [16], where PMMA samples were exposed to radiant fluxes up to 10 kW/m^2 , and the oxygen volume fraction for extinction (i.e., the amount of nitrogen added to air) for extinction was determined. Using a similar apparatus, Tewarson and Khan [52,53] determined the amount of Halon 1301 required to suppress a PMMA sample exposed to an external radiant flux of 50 kW/m^2. Taking a different approach, but again adding controlled amounts of energy to a polymer, Niemann and co-workers [9,54-56], and Driscoll and Rivers [57], describe tests in which a polymer sample is wrapped with Nichrome wire, which adds energy while the polymer burns. The extinguishing concentration with the added energy was determined for various suppressants. In a test procedure similar to the FM Global Fire Propagation Apparatus work, but with smaller samples, National Institute of Standards and Technology (NIST) workers [10,11], and Smith and Rivers [58] studied the effect of externally applied radiant heat on the suppression of PMMA. Since these tests (especially radiant heating) vary an important parameter, they could be used in a standard test method that is prescriptive. Alternatively, by specifying a particular energy input level, they could also be used for a prescriptive code.

Other tests which are of interest but which don't fit the above categories well are described below in the section: Other Miscellaneous Test Approaches. The hot-metal-surface autoignition tests of premixed fuel-air-suppressant mixtures, conducted by Hamins and Borthwick [12] and Braun et al. [13] describe the effects of the suppressant on the autoignition temperature. They also present the concentration of various agents necessary to inert a propane-air mixture to the hot surface ignition. Since these hot-surface tests study autoignition rather than suppression (which has different chemistry), as well as premixed flames (instead of flames over decomposing polymers) the results are not directly applicable, but are still of value to the present analysis. (For example, they study the re-light propensity of the gas-phase decomposition products in the presence of a hot surface, which may be relevant, but only for special situations.)

Recent National Aeronautics and Space Administration (NASA) tests, using three different models of damaged-wire ignition, evaluated the current necessary for hot-wire ignition of space suit materials [59]. Test data were presented for a range of wire sizes, polymeric materials, and surface conditions. Other tests at NASA studied the suppression of fires over resistively-heated PMMA cylinders [44,45].

Researchers at Technical Research Centre of Finland (VTT) studied mechanisms of electrical ignition in data and power cables in nuclear power plants [50]. Starting with statistical data on fire events, they categorized the ignition mechanisms. They then developed analytical models of the phenomena, and conducted supporting experiments to validate the calculations. While the VTT work might be considered to fall into the first category of test method (Tests Simulating the Failure Mechanism), the work studied ignition rather than suppression, so their results are not directly usable in the present work (although they are still of interest).

The above research is described in more detail below. The strengths and weaknesses of each test are outlined, and the experiments are analyzed to estimate the order-of-magnitude of the relevant parameters in each scenario (for example, the imposed heat flux), so that the results can be inter-compared.

Tests Simulating the Failure Mechanism with Suppression

The first test outlined in this section is the FM Global Electrical Arc Apparatus, since it was developed the earliest. Most of the other tests are based on work conducted or coordinated by Robin, and extended by Niemann and co-workers.

FM Global Electrical Arc Apparatus

In some of the first work to look at the suppression of simulated electrically-energized fires, Tewarson and Khan [52,53] describe an experiment with suppression of a fire sustained by simulated electrical arcing. The FM Global Electrical Arc Apparatus exposed burning copper-coated carbon rods, which were undergoing high-energy electrical discharge, to an atmosphere of air and Halon 1301. Power dissipated in the arc varied from 0.6 kW to 1.4 kW, depending upon the halon concentration and time (i.e., arc separation). The test chamber was a cube, 30.5 cm on edge, and the halon was introduced into the sealed chamber where a small fan mixed the agent (and remained running during the tests). The power to the arc was supplied by DC arc welder. It was found that an agent volume fraction of between 0.075 and 0.09 was required to extinguish the gas-phase flame of the carbon (from the copper-coated rods) with air.

In related work using the same apparatus, Khan [60] described tests in which the copper-coated carbon rods were covered with PVC cable insulation, and the concentration of Halon 1301 required to extinguish the flames over the PVC were determined. To establish the arc, the voltage was initially set to 50 V with a current setting of 40 A; however, these dropped off as the experiment continued, such that power levels to the arc were generally between 0.5 kW and 1.4 kW. A Halon 1301 volume fraction of 0.03 was found sufficient for extinguishment, although lower concentrations were not tested. As reported by Khan [60], the oxygen volume fraction in the chamber was measured, and dropped from 0.209 at the start to 0.19 after Halon 1301 was added to a volume fraction of 0.09. This drop represents the dilution of the chamber air by the halon and, hence, oxygen depletion due to combustion did not contribute to the extinguishment. As noted by Khan, this generally needs to be considered in sealed chambers.

The test method has the advantages of relatively well defined suppressant concentrations and flow fields and a simple configuration, and the test can be operated with a wide range of materials. The air currents in the chamber from the mixing fan could affect the stabilization conditions of the flame attached to the burning PVC on the electrode; however, the heating-induced flow from the arc near the PVC is probably much greater, dominating the stabilization conditions. The energy input to the arc is well defined, but that reaching the burning polymer is less well defined (since the polymer regresses away from the arc as it burns). Nonetheless, the test can serve as an upper limit of the effect of energy addition, since flames extinguished near the arc discharge are probably well-stabilized, exposed to a high radiant (and conductive) heat flux, and have a continuous ignition source; i.e., a worst-case scenario.

Failure Mechanism-Based Approach of Robin, McKenna, Stilwell, and co-workers

McKenna and co-workers [4,5] reported a series of tests which aimed to simulate suppression of electrically-energized fires in central office equipment. Three test configurations were conceived, built, and tested, and a range of materials were used. The goal was to replicate both the electrical involvement in the fire (with regard to both ignition sources and energy addition), as well as the materials and suppressants used in practice.

Three configurations were devised: 1.) the Ohmic Heating Test; 2.) the Over-Heated Connector Test; and 3.) the Printed Wiring Board Test. The first simulates an over-heated cable fire, while the second simulates an overheated connection, each of which could occur from a shorted connection in energized electrical equipment. The third test simulates the development of an arc-track leading to a continuous arc between parallel power circuits on a printed wiring board. These three tests are very useful for understanding the general behavior of the burning polymers when the electrical power in the circuit adds energy to the system, and represent some of the few quantitative tests of suppression of burning polymers heated by simulated electrical shorts. Nonetheless, there are some physical parameters in the tests which are crucial for understanding both the burning rate and the suppression characteristics, but that are either not controlled in the work, or not reported. The same physical enclosure and system for agent addition were used for all of the tests.

The common test enclosure has some properties which affect all three of the tests listed above, and hence is discussed first. For example, the agent addition is impulsive, which leads to a few complications: 1.) turbulent fluctuations in the local concentration of agent where the extinguishment is actually occurring, 2.) a rapidly changing average concentration in the time during which the extinguishment occurs, and 3.) non-uniform mixing of the agent with the ambient air, so that there can be spatial and temporal concentration gradients in the enclosure. These three effects make it difficult to know the actual concentration of agent at the burning polymer when extinguishment occurs. For example, in a test, the mass of agent in the bottle was set to provide a final concentration of agent in the enclosure corresponding to that with complete mixing of the agent with the air in the enclosure. When the agent is released, however, it forms a jet of agent with a concentration that is locally much higher than the final design concentration. The concentration of agent delivered to the fire is the fundamental parameter desired, but it is poorly characterized in this test. Note for example, Figure 6 in the McKenna 1998 report. As indicated in the figure, the average volume fraction (from a line-of-sight FTIR measurement) changes very rapidly at short times, going from 0 % to about 7 % (at the measurement location) in about 6.3 s. For comparison, the average fire out times are 9 s to 20 s for the conductive heating tests, 3 s to 16 s for the ohmic heating tests, and 2 s to 9 s for the printed wire board tests. Hence, the agent concentration is changing very rapidly in the same time scale of the extinguishment process, so that it is difficult to know what the concentration actually was when the fire went out. The situation is even more uncertain since the line-of-sight FTIR measurement spatially averages the values of the concentration, so that local fluctuations are likely to be greater. Also, the location of the burning polymer is even further along the flow streamlines than the FTIR measurement, so the time to reach peak concentration at the polymer will be even larger than the 6.3 s in the figure (which is the concentration at the FTIR-beam location). Since the agent release is impulsive and turbulent, the concentration at the burning polymer is probably also stochastic, further complicating data interpretation.

Another general concern has to do with airflow. The burning rate and stabilization (and hence, extinguishment conditions) of polymers can be sensitive to the air flow over the surface. Since most electronic equipment will have substantial cooling air flows, these must be considered in the test procedure. The airflow at the burning surface in the present test enclosure is uncharacterized, and that situation is made even more tenuous with the impulsive release of the agent, which could have small effects on the local flow field, that could affect the stabilization of the flame.

The stochastic nature of the concentration fluctuations and the oxidizer flow near the fire means that either a lower precision in the measurement must be accepted, or a much larger number of tests must be performed to provide an average value. While the test is perfectly reasonable for investigating the general behavior of a burning polymer in a configuration similar to that in telecom central office equipment, it is not as good for characterizing the behavior in terms of two of the most relevant parameters (e.g., extinguishment concentration and flame stabilization condition) .

Finally, since the test chamber is sealed, with the polymer still burning, prior to the release of the agent, the possibility of suppression under vitiated conditions (i.e., depleted oxygen) exists. Some care must be exercised to insure that oxygen levels at suppression are still at ambient conditions. Besides these generic properties associated with the test enclosure, the individual tests are discussed below.

Ohmic Heating Test

In the Ohmic Heating Test, bundles of cables ranging in diameter from 3.26 mm to 0.403 mm (8 AWG to 18 AWG), with jacket materials typically used in telecom (see Table 5), were arranged in bundles, and some fraction of them was heated with a high current/low voltage resistive source. The goal was to auto-ignite the jackets with the ohmic heating; however, this is found to be difficult to accomplish due to melting, dripping, or smoldering of the jackets, and failure of the conductor. Since the delicate balance necessary for self-ignition was difficult to achieve, a butane pilot flame (of 25 s to 170 s duration) was ultimately used for the initial ignition of the polymers (except for the non-FR polyethylene which self-ignited).

Table 5- Jacket materials use in work of McKenna et al. [4]

1. Cross-linked polyethylene (XLPE),
2. SJTW-A: Thermoplastic jacket over thermoplastic insulation,
3. Polyvinylchloride (PVC),
4. Chrome PVC jacket over polyethylene,
S. Neoprene jacket over rubber insulation.

These types of tests are very useful to start to get in the ballpark of how energy-augmented combustion (EAC) affects the minimum extinguishing value for suppressant agents, $X_{a,ext}$. As a research tool, they provide excellent insight into the behavior of the tested materials when subjected to simulated failure modes expected in the field. Nonetheless, the tests performed and the results provided did not give all of the relevant parameters necessary to characterize the behavior; i.e., the fundamentally controlling parameters were sometimes not provided. For example, the net power into the test material is what controls the polymer temperature (and hence the mass loss rate), but it was not provided in the reports. As described above, because of the delicate balance between heat feedback to the polymer, heat losses, and the fuel decomposition rate (which creates the fuel), the net power into the polymer has a big effect on the suppressant requirements. Also, near extinguishment, the burning rate (and hence the required suppressant concentration for extinguishment) can be highly non-linear with the input power, so it is important to report the extinction concentration of agent as a function of the input power level. A second concern has to do with the pilot flame used for ignition. Since the resistive heat source creates a polymer temperature field that is time-dependent, the burning behavior (and hence suppressant requirement) can vary widely with the time history of the heating. That is, the polymer can be much easier to put out early in its heating history, but more difficult later when the heat release will have grown geometrically. It may be more appropriate, if an external ignition source were used, to keep it on for the duration of the test. (Alternatively, one must make arguments about *when* in the heating history to light the polymer, and for how long to let it burn, before suppression).

Also, it is important *where* in the polymer the energy addition is made. In some cases, the energy from the heated cable is added to the center of a bundle of cables, which is probably not where the cable is burning. If the sample is thermally thick, and the energy is added to the back side (or the center) of the sample, the energy will have a smaller effect on the burning rate (and the MEC) than if the energy is added at the surface. Further, the burning rate depends not just on the total energy added, but also on the energy density. That is, adding 20 kW/m^2 to a single area has a much different effect than adding 10 kW/m^2 to twice the area. While failing, burning cables could conceivably be in any configuration, the most conservative case would be for the energy to be added at the burning surface. This is not always the case in the present study.

Many of the tests here picked a particular configuration and power level, and then determined the agent concentration required for suppression. Since the relevant parameters in an actual electrically-energized fire scenario are not that well known, it would be very informative to run the present tests for a range of power levels, air flow conditions, polymer containment conditions, and flame heat feedback conditions, to see the effect of these parameters on the MEC.

Overheated Connection Test (Conductive Heating Test)

In the overheated connection test, a cylindrically-shaped heater (a "ring heater") clamped onto and heated one end of a relatively large diameter power conductor (350 MCM, or about 17.3 mm diameter). The clamped end of the wire (about 10 cm long) was stripped of the polymer insulation, while about 15 cm of insulated cable extended vertically above the heater. The heater temperature was set to 900 °C, and it heated the cable until the far end of the cable reached 310 °C, when a pilot flame was applied to the base of the exposed insulation for 15 s. The agent was injected, and the time to flame extinguishment was recorded.

As with the ohmic heating test described above, this test is a very reasonable approach for starting to understand the behavior (and suppression) of burning insulation on real power conductors exposed to heat loads when they are arranged vertically in isolation. As with the other test, additional information concerning the test conditions could make the test more broadly useful. (For example, the temperature of the polymer surface, the wire temperature along the length, and the power going into the wire itself would be of value.) Concerns with this test are listed below.

Since the same enclosure is used for this test as in the ohmic heating test, all of the issues discussed above with regard to the agent addition and mixing apply here as well. As with the other test, it would be nice to include measurements or estimates of the power addition to the burning material, and to have performed the tests at varying power levels. The airflow velocity and configuration could have been better characterized, since these affect the flame stabilization and hence, the blow-off condition.

It is of value to know the energy flux to the polymer so that it can be compared with other test methods. The external heat flux can be roughly estimated as follows. The power level to the ring heater was 1 kW. For a copper rod (assumed solid core, and a thermal conductivity of 0.36 kW/m/K) with a linear temperature distribution and 900 °C at one end and 310 °C at the other, and 15 cm in length, in steady-state, heat is conducted into the rod at 0.33 kW. This value is about 1/3 of the total heat to the ring heater, which seems reasonable. The wire has an external surface area of about 80 cm^2, so the average heat flux into the polymer insulation estimated here is 45 kW/m^2. The heat loss through the polymer jacket along the wire varies significantly, because of the temperature gradient in the cable. Nonetheless, the temperature gradient along the length of the wire is likely steeper near the ring heater than at the far end (because the higher temperature at the ring end leads to larger heat losses than at the cooler end). This would make the power dissipated in the wire larger than the estimate here (by assuming a linear temperature distribution); the *variation* in the energy flux would remain the same, however, since the variation is maximum between the end points of the wire, and these are the same regardless of the temperature profile between them.

The variation in the heat losses along the cable are estimated as follows. The insulation must be considered, but there are two limiting cases which can bound the problem: no insulation, and insulation with a constant temperature (corresponding to a constant decomposition temperature of the polymer). For the first case, neglecting the polymer insulation, the heat losses by convection are linear with the temperature, so they vary by a factor of three along the cable, while the radiative losses (which are proportional to T^4) vary by about a factor of 18. Assuming an emissivity of 0.95 for the wire (dirty, carbon coated), and a convective heat transfer coefficient of 10 W/m^2/K, the estimated heat loss from the wire (which is equal to the heat flux into the wire insulation at that location) is shown in , and varies from 8.1 kW/m^2 to 107 kW/m^2. The average heat loss in the wire is 45 kW/m^2, (which is the same as the value above calculated from the heat conduction input through the wire cross-section). For the second case, constant polymer temperature, the heat losses are only by conduction to the polymer and depend only upon the polymer temperature and the cable temperature (for the limiting cases, we assume here that there is complete absorption of the thermal radiation from the wire by the polymer).

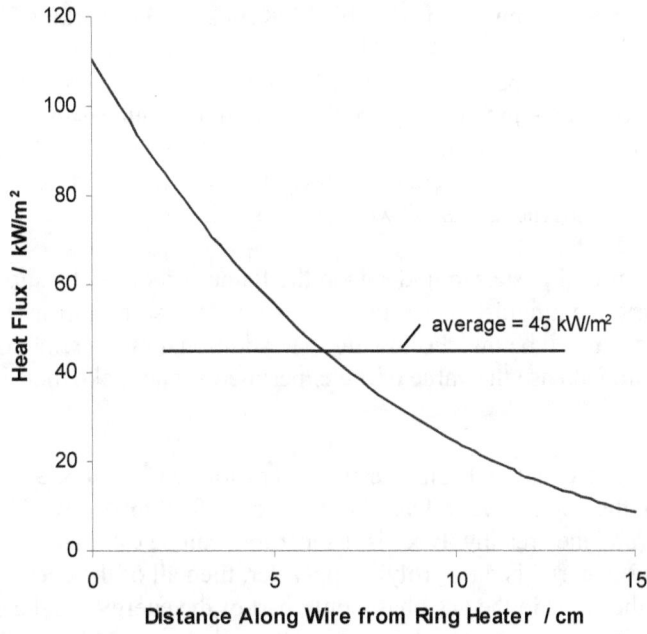

Figure 15 – Estimated heat flux along wire in heated conductor test [5].

33

The polymer likely decomposes between 300 °C and 500 °C, and in either case, the heat losses out of the cable vary along its length by even more than they do in the case of no insulation. (This is because at either polymer temperature, the heat flow near the colder end of the cable is near zero, but is a larger, finite value new the hotter end.) Using the first case (no insulation, which predicts less variation in the heat flux along the wire), the heat flux into to polymer is likely is in the range of 8 kW/m^2 to 110 kW/m^2. The absolute power levels are probably greater than this since the heat losses are not constant along the rod so that the temperature gradient in the rod is not constant, but rather, steeper at the hot end.

Modified Conductive Heating Test

Niemann, Bengtson, Flamm, and Smith [14,15,51] describe a modified version of the conductive heating test of McKenna et al. described above. In it, they add a spark ignition source to the top (i.e., coolest part) of the heated cable, and find that higher concentrations of agent are required to suppress all forms of flaming combustion. Since the conductive heating test of McKenna et al.[4] also uses a separate ignition source, there is some logic to keeping the ignition source in place for the duration of the test. In an actual fire scenario, if the ignition source—independent of the continuous cable heating—existed at some early time in the fire development, it can also exist at later times, so inclusion of a continuous ignition source, as suggested by Niemann et al. seems reasonable. On the other hand, their criterion for flame extinguishment may be too stringent. Flames existing *only* near the arc igniter (of these tests) may be due to the localized energy release of the arc, which is a separate phenomenon from what is intended by the ring heater (the energy density near the plasma of the arc is very high). Finally, the most likely region to support a flame is at the hot end of the rod, where the external heat flux is greatest (see), and where the flame stabilization is strongest, due to the downstream boundary layer [43]. Hence, while their criterion for flame extinguishment may be too stringent, the location of their spark igniter may not be the most conservative.

Printed Wiring Board (PWB) Failure

In a series of tests with printed wiring boards (PWB), McKenna and co-workers [4,5] simulated an arcing fault across power tracks. Varying the track width (0.304, 0.63, and 1.27) mm and track spacing (0.304, 0.63, and 1.27) mm, solder mask type (LP1 or Vacra 1), and substrate material (FR-4 or FR-2), they attempted to obtain the most stable propagating arc faults for track voltages of 5 V to 9 V, with a current of 11.5 A. After the optimal conditions for a stable arc were established (8.5 A, 8.75 V), they attempted to extinguish flames on the board which were self-ignited by the arc. Quantities of HFC-227ea necessary to extinguish the flames (but not the arc itself) were established.

The experiments are very useful to start to understand the flames produced by such failures, as well as the amount of agent necessary to extinguish these flames. Using the descriptions in the reports, the energy flux to the substrate are estimated below. Following that, additional measurements, tests, and analyses are suggested which would increase the value of the experiments, and make their interpretation more universal.

The energy flux to the substrate material can be estimated as follows. With the given current (11.5 A) and voltage (5 V to 9 V), the power level released in the arc is 58 W to 104 W. The fraction of this power making it to the surface (and the area involved) is unclear, but can be estimated. If the electrically conducting (i.e., shorting) material is the pyrolyzed polymer, then all of the energy will be delivered to the polymer; whereas if the arc is in the gas phase, only part of the energy will be delivered. Assuming the second case, a lower limit of the energy delivered to the polymer would be about 50 % (radiation and conduction are about equal towards and away from the surface). As for the area involved, the flame is described as about 2.54 cm in width, and assuming the depth is the same, we get an involved surface area of 6.5 cm^2. This estimate of the area is reasonable since the flame has a large effect on the polymer surface, promoting carbonization (and hence shorting), as well as on the gas-phase arcing (where ions and carbon in the flame again promote shorting through the flame region). If anything, the arc-heated area may be smaller than the flame area, since conduction within the substrate from the high-temperature arc-heated area will cause mass loss and flaming from a larger area than that of the arc heating. Assuming half of the energy into the wires makes it into the polymer, and an area of 6.5 cm^2 gives fluxes of 45 kW/m^2 to 80 kW/m^2, for 5 V and 9 V, respectively. (Note, the other publication of this same work [4] describes the power input as 74 W, which with the above assumptions, gives about 57 kW/m^2.)

In order to extend the value of the PWB tests, several additional actions could be taken. For these tests, the effect of a second board in close proximity to the first would be very interesting to examine. While the authors did describe tests in which the arc was initiated on one board, with a second board close to and parallel to the first, they only noted whether the flame initiated on the first board then propagated to the second board. This is a useful result, but more interesting would be whether the flame on the second board affected the extinguishment of the flame on the first board—since heat feedback from adjacent flames is known to affect the heat flux to and burning rate of flames over condensed-phase materials [61]. Further, PWB in racks have ventilation air from cooling fans, and the average flow velocity over an individual board likely varies significantly. Since the burning rate (and the flame stabilization properties) vary with the airflow [43], it would be useful to see how the extinguishing concentrations vary with airflow velocity (and direction). The effects of ventilation are particularly important for parallel boards in close proximity, since under those conditions, the burning in the central region is likely to be ventilation limited. The effects of board orientation are a step in the direction of understanding some of the ventilation effects; however, the orientation effects were mentioned in the reports, presumably only for single boards, and no results were given.

Vertical Polymer Slabs Ignited by a Loop of Nichrome Wire

Robin, Shaw, and Stilwell [6] devised a test for assessing the effects of energy-augmented combustion on the clean-agent suppression of burning polymer samples. The configuration, shown in , uses a U-shaped length of Nichrome wire which passes through rectangular slots in a vertical polymer slab. The Nichrome wire is resistively heated to provide a desired wire temperature. In these tests, the temperature is initially set to 1256 K (1800 °F) for the first 30 s (to establish ignition and burning of the polymer), followed by a setting of 922 K (1200 °F) for the next 30 s. The apparatus in is placed in a large (1 m x 2.3 m x 2.4 m) enclosure, with a single nozzle for impulsive release of the suppressant agent, and baffling similar to the UL2166 test. The suppressant is introduced to the enclosure at 60 s.

Figure 16 – Test method employing U-shaped Nichrome wire in proximity to a vertical polymer sample (from ref [6], used with permission of author) .

The test configuration has similarities to hot wires in contact with polymers, which might occur in electrically-energized telecom or data processing equipment. It has the advantage of variable wire temperature and polymer type, allowing examination the sensitivity of the suppression process to these variables (although only data for variation in the polymer type was reported in ref. [6]). The tests would be more broadly useful if the power input to the Nichrome wire were given, and if the fraction of the total power which goes into the polymer were provided as well. The minimum concentration of agent for extinguishment for a given power level would be very useful (as opposed to a pass/fail result for one agent at an unspecified power level). Also, the distance from the wire to the polymer sample when the suppressant is added would be helpful for characterizing the heat flux to the polymer. To some extent, the test procedure is like other polymer burning tests with radiant heat addition, with a difference here being that the radiant source is a small wire at higher temperature (rather than a cone heater [11] or quartz heaters at lower temperature [16,53]), with a radiation intensity which has not been characterized. In addition to the radiant and convective heating of the polymer, the wire also provides an ignition source. Rather than attempting to obtain a simplified 1-d burning configuration as in the other tests, the present test has a complicated 3-d configuration with regions of electrical heating different from the regions of flame heating (since the flames extend up the sides of the vertical polymer, and are probably not present in the small, flame-quenched area in the slot where the wire passes through).

This test is part way between a simulation of some failure mode in the field, and a model, well-controlled experiment controlling a single parameter. While the test method has some positive attributes and potential, it also has some shortcomings. The discussion above with regard to the unsteady, impulsive flow of the suppressant agent (with a mixing time constant of the same order as the extinction time) is valid here. It would be more tractable to place the sample holder in into a flow tunnel with steady, well characterized flow (and pseudo-steady agent concentration, slowly increased), so that both the actual agent concentration (and the flow field) were better known at the extinguishment condition. A major shortcoming of the method is that the actual heat flux from the wire to the polymer is both unknown and changing with time as the polymer surface regresses from the wire. Also, the heat flux from the flame is difficult to estimate since it is not clear what part of the flame (if any) provides heat feedback to the region with the electrical heating. Estimates of the time-varying heat flux are described below. Finally, it is the net heat flux to the pyrolyzing region of the polymer which leads to mass loss. As described above in the Literature Review section, a major difference between the net and gross heat flux into the pyrolyzing region is due to conductive losses (into the polymer), and this changes with time, particularly at short times, and for three-dimensional configurations.

To compare the present test to other tests with added energy, it is of interest to estimate the heat flux from the hot wire to the polymer. Modeling the wire as a horizontal cylinder, it is possible to estimate the radiative, conductive, and convective heat losses from the wire [62,63]. Assumptions in the calculations are as follows. For radiation, the calculation is straightforward, with the only variables the wire temperature (known), the surface emissivity, and the distance of the wire to the polymer surface. The emissivity of the Nichrome wire was assumed to be unity (values range from 0.79 for bright wire, to 0.98 for oxidized wire [64], while if the wire were dirty due to polymer residual, the number would be near 0.95). For simplicity, the emissivity and absorptivity of the PMMA to IR radiation was also assumed to be unity [18]). The flux on the PMMA was calculated from the net energy radiated from the wire to a PMMA cylinder of radius equal to the separation distance plus the wire radius. This radiative heat flux to the polymer as a function of separation distance between the wire and the polymer is shown in Figure 17 (dot-dashed line).

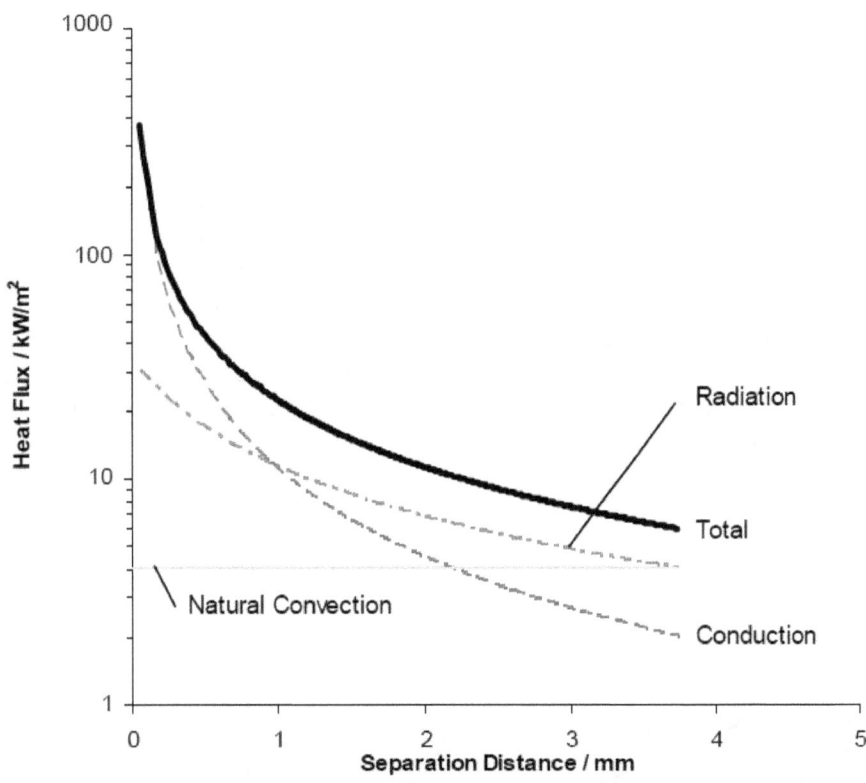

Figure 17 – Estimated heat flux from a wire at 922 K (1200 °F) to a polymer (at 330 °C), as a function of their separation distance.

The conductive/convective heat flux from the wire is more difficult to estimate since the flow configuration is not simple. Two limits can be obtained by assuming: 1.) free, natural convection from a horizontal wire, 2.) conductive heat loss from concentric cylinders (i.e., the wire with a concentric polymer surrounding it, separated by an annular region of air). Since the wire is initially very close to the polymer, free convective flow is not possible, and domain-limited conduction is more realistic. Nonetheless, the actual conductive layer will be complicated by blowing (mass loss from the polymer), which will tend to reduce the conductive heat flow to the polymer. As the separation becomes larger than the boundary layer, natural convection can develop; the separation distance of the polymer from the wire at 60 s is of interest, but was not provided in the report. For the conductive heat flux, the standard concentric cylinder estimate for heat flow was used [63], with air properties at the mean of the surface temperatures. For the free convection, the standard correlations for a cylinder in cross flow (ambient air at 298 K) provided the heat losses [63], and all of the heat was assumed to impinge on the top of the slot in the PMMA (which has an area of 0.51 cm^2).

Estimates of the heat transfer from the wire to the top surface of the polymer slot (see) from pure conduction and free convection are also shown (dashed and thin solid lines), as is the total heat flux (black line). The total includes the radiation and pure conduction, since the latter is more likely than free convection for the small separation distances expected. As indicted in the figure, there is large variation in the total heat flux with separation distance, from about 40 kW/m^2 at 0.5 mm separation, to about 10 kW/m^2 at 2 mm separation. The average heat flux will vary with polymer type since their regression rates (and hence, separation distances), will vary. Note that in general, very large variations in the heat flux can occur as the polymer regress, from about 100 kW/m^2 at 0.33 mm separation, to 6 kW/m^2 at 4 mm).

Using the heat flux predictions in Figure 17, together with the heat of gasification (1600 kJ/kg) and density (1200 kg/m^3) for PMMA [65], it is possible to estimate the regression rate of the PMMA (neglecting flame heat feedback), and then plot the separation distance (and heat fluxes) as a function of time, for the PMMA in the test as conducted. This is done in ; note that at t=30 s, the wire temperature changes from 1256 K to 922 K, so the heat fluxes also change at t=30 s. The local heat fluxes are very high at low times, but decrease rapidly as the surface regresses. The estimates here may also be higher than in practice due to neglect of the conductive losses into polymer. These losses create lower net heat flux into the polymer, lowering the mass loss rates at early times—before a steady-state temperature distribution in the polymer has been established. Nonetheless, the three-dimensional, irregular, shape of the sample and non-uniform heat flux make estimates of the conductive heat losses into the sample difficult.

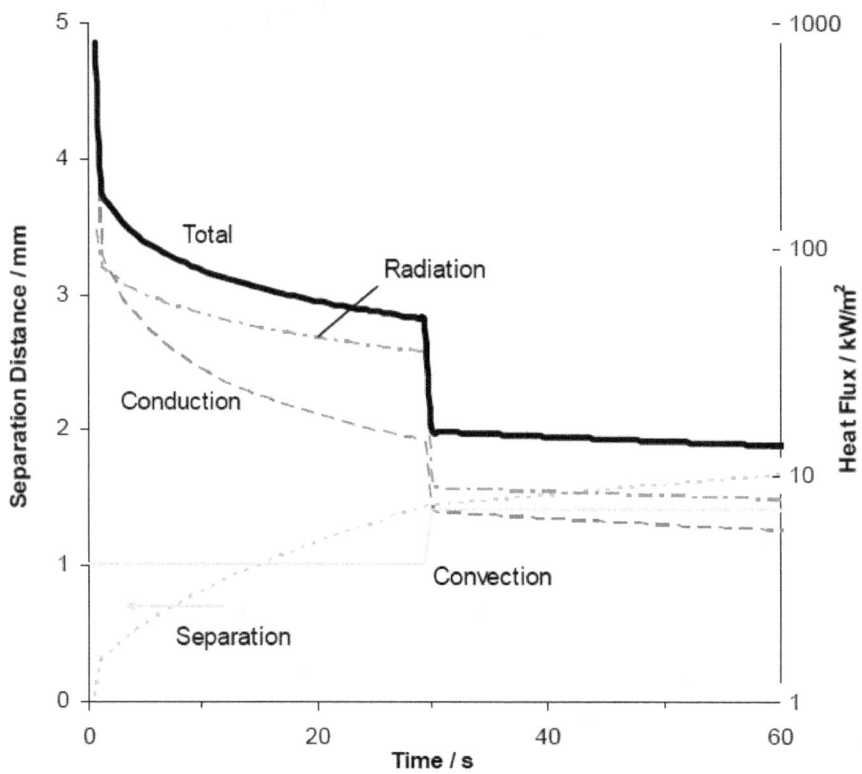

Figure 18 – Estimated separation distance and heat fluxes (radiation, conduction, and free convection) as a function of time for a wire close to a horizontal PMMA surface.

Despite the challenges in completely quantifying the conditions of the test which would be useful for its use as a test method, the experiments have tremendous value. As the calculations above indicate, in this experiment designed to simulate a failing, yet still energized electrical component, the estimated heat fluxes to the surface are in the range of 6 kW/m^2 to 100 kW/m^2. Hence, for experiments which more carefully specify the external heat flux, values in this range could conceivably be relevant.

As indicated in , the estimated separation distance at 60 s is about 2 mm. The actual separation distance in the experiment, however, can vary due to: experimental positioning errors, sagging of the Nichrome wire when heated, flame-induced heat transfer, and buoyancy- and capillary-induced flow in the polymer melt [46]. The purpose of the heat flux calculation above is not to predict, a prior, the separation distance as a function of time, but rather, to show the *variation* in the heat transfer with separation distance, so that the difficulty of accurately controlling this parameter in the test is illustrated.

Finally, the area for heat addition from the electrical source is not the same as the area for flame heat feedback. This is because the flame extends up the side of the PMMA sample, rather than being attached to the underside of the slot where the wire is adding its heat. The result is that the total heat input into the polymer is distributed to a larger area, so that the conductive heat losses—which are very important at short times—become very large, leading to a lower burning rate and weaker flame than if the electrical energy and the flame heat flux were added at the same surface. Also, because of the complex configuration, the heat feedback from the flame to the surface is poorly characterized and hard to estimate.

It would be valuable to compare the extinguishing concentration of HFC-227 in this test to that in other configurations which involve energy-augmented combustion. To do this we need to know the heat feedback from the flame to the polymer, the conductive heat losses to the interior of the polymer sample at the time of extinguishment, the heat flux from the wire to the polymer. These can be estimated, but would require significant work. The mass loss rate data as well as the MEC as a function of wire temperature and preheating time would be very informative.

Wire Cable Bundles with Heated Nichrome Wire

In a follow-up conference paper, Robin et al. [7] describe a test to simulate suppression of energy-augmented combustion fires. In the test, an assembly of seven wire cables, each 15.2 cm long, is grouped together. The jacket of each cable contains a number (perhaps five) of individual insulated wires (unspecified size). An 18 AWG (1.024 mm diameter) Nichrome wire is inserted into the jacket of the central-most cable, and the Nichrome wire is heated to 1800 °F (982 °C). The wire ignites and burns for 60 s, when the suppressant is added. The same agent injection system and enclosure is used as in the Vertical Polymer Slab test described above. While the test does assess the suppressant requirement for HFC-227 for this particular configuration, it is not a very challenging test, and few details are provided which would help to make the results more universally useful (for example, the power input to the wires is not supplied, so the energy flux cannot be estimated).

The main difficulties have to do with the short time scale of the test, the area where heat is applied, and the flame stabilization. The heat is applied deep within the cable bundle, whereas burning occurs from jets of fuel gases from the pyrolyzed polymer insulation emanating from the ends of the cable bundle where the Nichrome wire enters and exist. Hence, the flames likely extend several inches up past the burning cable bundle, where they provide little heat feedback to the polymer, and even less (or no) heat to the area where the resistive heat is applied. Because the heated wire is buried deep in the cable bundle, most of the energy is spread out to metal and polymer mass which is not participating in the combustion process. That is, the cable bundle act as a large heat sink, so that most of the energy being put into the system preheats the insulation and wire, but does not cause more mass of polymer to be pyrolyzed. If the sample were allowed to burn longer, the burning rates (and likely, required suppressant concentrations) would be higher. In the present test, most of the energy conducted into the sample from the hot wire preheats the mass (which has not yet had a chance to burn). Placing the Nichrome wires on the outside of the wire bundle (and set to some reasonable power level) would be a more challenging test configuration for the suppressant, yet is still a plausible scenario.

Test Methods Based on Controlling the External Heat Flux

Other researchers have taken a different approach, and have devised experiments in which the heat flux to the burning polymer is more accurately specified and controlled. These include tests in which the polymer is heated (externally or internally) by resistive Nichrome heating wires [9,54-57], and others which impose a radiant flux on the sample [10,11,16,52,53,58]. These researchers, while demonstrating the effect of the added energy on the suppressant requirements, typically have made no effort to estimate the appropriate external flux level to simulate suppression of actual failure modes in electrically-energized fires.

Resistively Heated Polymer Samples

Niemann, Bayless and Craft [9] report a test method for the suppression of resistively-heated polymer samples. In it, the test sample (in this case, PMMA) is heated with Nichrome wire, which is either wrapped on the exterior surface, or sandwiched (with spacers) between two slabs of the polymer. The polymer sample is placed in a V-shaped holder, which is centrally located and raised about 20 cm above the floor in a test chamber (measuring approximately 1 m on each side). The suppressant agent is added to the enclosure with a single nozzle located near the top, and injection velocity, together with buoyancy-induced natural convection currents (from the burning material) in the enclosure, provide mixing of the agent with the air. Two power levels were tested: 48 W and 192 W, and the test concentration necessary for extinguishment was determined.

As discussed in the section above, the unsteady agent addition and poorly characterized agent mixing with the air lead to two problems in data interpretation. First, it is difficult to know the concentration of agent (or it's time variation) actually reaching the burning polymer when it does (or does not) extinguish. Second, the stochastic nature of the mixing and release process will lead to natural variability in the test results, requiring a larger number of tests to accurately define the concentration boundaries for extinguishment, or non-extinguishment. Also, when the agent is released impulsively, it is hard to interpret the effect of flow field changes due to the impulsive agent release on the flame stabilization. That is, the flame stabilization may be modified by subtle changes in the flow field near the stabilization point of the attached flame on the polymer when the agent is impulsively added to the test volume.

Despite the shortcomings in the method, it has some advantages. It does provide a general overview of the effects of heat addition to a burning polymer on the MEC. As with other tests with Nichrome wire described above, the amount of added energy *could* be continuously varied, so that suppressant requirements could be determined as a function of added power (although only two power levels were used in the present work). The heat is added to the polymer surface (in the 192 W case), which is where it can have the largest effect on the suppressant concentration for extinguishment. Also, the heat feedback from the flame overlaps to some extent (but not completely) with the area for heat addition from the wire. As with most of the other tests described above, any material or agent can be tested.

A challenge in interpreting the test results concerns the amount of energy added to the polymer. From the physical arrangement of the polymer sample with heating wires, it is difficult to know what fraction of the energy added to the wires actually makes it into the polymer. This is especially true for the configuration of the 192 W heat input case, but also true to some extent for the 48 W case (i.e., what fraction of the wires is in the polymer vs. outside). Assuming that 100 % of the energy in the 48 W case makes it into the polymer (which sandwiches the Nichrome wire), this power corresponds to an energy flux of 6.8 kW/m^2. For the 192 W case, the energy flux is 110 kW/m^2 to 220 kW/m^2, based on 50 % or 100 % of the energy making it into the polymer.

Driscoll and Rivers [57] report further results using the same apparatus as in Niemann et al. In Driscoll and River's work, several shapes of PMMA are tested, and a cylinder (7.6 cm length x 2.54 cm) diameter was chosen for further testing. This cylinder was wrapped with Nichrome wire, and 225 W was added from a 12 V source. The energy flux to the PMMA was 32 kW/m^2, based on the total energy to the wire, or 16 kW/m^2 based on half of this energy making it inward). The concentration of HFC-218, HFC-3-1-10, HFC-23, and HFC-227 necessary for extinguishment (and to prevent re-ignition) was measured (along with acid gas production). Niemann and co-workers [55,56] continued the work in ref. [54] to provide data for extinguishment by HFC-125, HFC-218, and FK-5-1-12.

It would be useful in tests such as these for the researchers to always include data on the suppressant concentrations with no heat addition (or better, as a function of power levels). Then, the quantitative effects of the heat addition on the suppressant concentrations would be clearly demonstrated. Nonetheless, the results for the resistive wire heating of polymer samples can be compared with other tests with added energy, and this is done in the section "Equivalence of Radiant and Conductive Heat Addition" below.

FM Global Fire Propagation Apparatus

In early work describing test methods for assessing the flammability of plastics, Tewarson and Pion [16] describe the equivalence of increased oxygen mole fraction in the air with externally added radiant heat. Using their method, they estimate the heat flux from the flame to the polymer, heat losses from the polymer surface, heat of gasification/pyrolysis/depolymerization, and the "ideal" burning rate (which is defined as the ratio of the heat supplied by the flame to the heat required to gasify/de-polymerize/pyrolyze the polymer; i.e., the burning rate which would occur if any heat loss terms were matched by an external heat input). Their method uses a horizontal polymer sample 60 cm^2 to 100 cm^2 in a chimney, with controlled atmosphere, and exposed to an external radiant heat flux. As the authors suggest in the conclusions of the paper, suppressants could be added to the air stream. While the authors did not present the data in such a way, it is possible to extract the values of the volume fraction of nitrogen necessary for flame extinguishment of the PMMA samples, as a function of the external radiant flux. shows the effect of external radiation on the required nitrogen for extinction of flames over PMMA in air from their data. As indicated, an external heat input of around 7 kW/m^2 leads to a doubling in the MEC. These tests are conceptually the same as those in the Radiantly Enhanced Extinguishing Device (REED) described below, but the sample is larger, and the experiments in ref. [16] predate those in the REED by twenty years. Although Tewarson and Pion allude to suppression tests in the device with other agents, they do not perform those tests in this work.

FM Global 50 kW-Scale Apparatus

The FM Global 50 kW-Scale Apparatus of Tewarson and Khan [52,53] is used to expose 10 cm x 10 cm x 2.54 cm PMMA slabs to radiant fluxes of 50 kW/m^2. A chimney permits addition of controlled atmospheres containing increasing amounts of suppressant (in these tests, CF_3Br), until extinguishment is achieved. The experiment measures mass loss rate, heat release rate, combustion efficiency, and production rates of CO, unburned hydrocarbons, and fluoride and bromide ions. At 50 kW/m^2, a volume fraction CF_3Br of about 0.06 is required to extinguish the PMMA, which is higher than the results of Bayless et al. [57] (0.03 and 0.048 ± 0.008), obtained at lower input energies (6.8 kW/m^2 and 16 kW/m^2 ± 5 kW/m^2).

Advantages of the test are a relatively well defined suppressant concentration and flow field, simple configuration, the possibility of testing a wide range of materials, and a relatively well defined energy input.

41

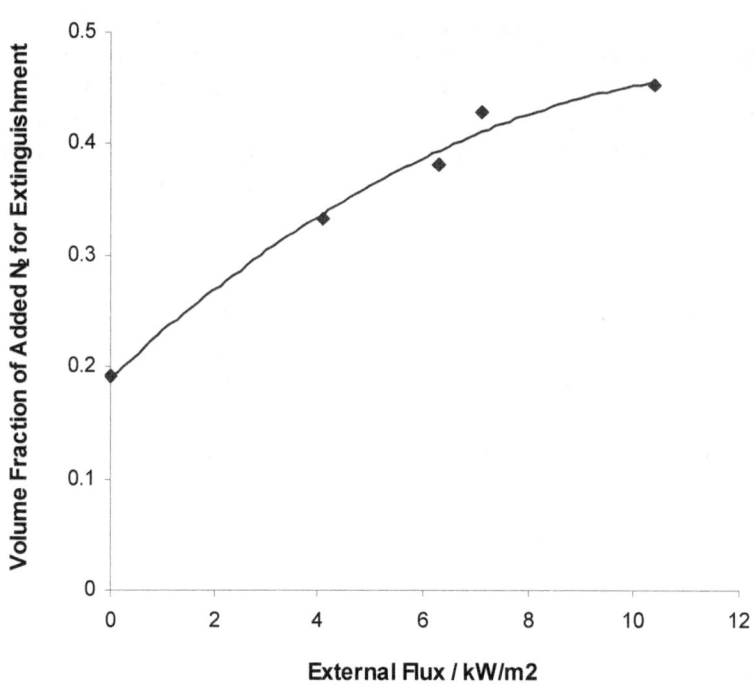

Figure 19 – Required nitrogen added to air for extinction of flames over PMMA as a function of external radiant flux [16] (points: experimental data; line: curve fit).

Radiantly Enhanced Extinguishing Device:

Steckler et al. [10] and Donnelly and Grosshandler [11] describe a device for estimating the effects of external heat loads on the clean-agent extinguishing concentration for flames over burning polymers. The REED is a cross between a cup burner and the cone calorimeter. In it, a small, 2.54 cm diameter, 2.54 cm long polymer cylinder is placed at the location of the usual fuel cup in the cup burner, and air flows up a surrounding chimney, with co-flowing guard nitrogen in a second chimney around the first. A radiant heater identical to that used on the NIST cone calorimeter [66] sits above the polymer sample and provides a known radiant heat flux to the burning polymer sample. A propane torch lights the polymer sample, which is pre-burned for 200 s. Agent is added to the air stream incrementally, with a 30 s waiting period between concentration increases. The procedure is repeated until extinguishment.

The technique has some desirable properties. The concentration of agent at extinguishment is well known, as is the radiant flux applied to the surface. The stabilization conditions of the attached flame are nearly constant and well-characterized. Any material which can be formed to the required sample shape can be tested, and the extinguishing concentration can be examined for a range of imposed external heat fluxes (from 0 kW/m^2 to about 150 kW/m^2).

The method also has some shortcomings related to the net energy transferred to the sample. The heat flux delivered to the sample depends upon the absorptivity of the sample to infrared radiation from the cone, and for some materials, this can change somewhat during the test. Also, for charring materials, a char layer will act as an insulator, and reduce the net energy conducted to the decomposition layer. While this isn't necessarily bad (the same polymer subjected to an electrical resistive heating load will act similarly), it requires that some care be exercised when designing the test protocol and data reduction, so that comparisons of required suppressant concentrations are made under consistent net heat flow conditions. Similarly, heat losses due to conduction into the interior of the polymer sample will reduce the net energy available for fuel-species generation, and these will be unsteady in time. The temperature profile in the solid polymer changes with time, especially at short times, and these are somewhat difficult to characterize because of three-dimensional effects for the small samples of the test. The effects of varying conductive losses play the same role as variable pre-heating of the sample, which has been shown to affect suppressant requirements for burning polymers [44,45], as well as the electrical ignition of PVC cable [67]. Hence, it is important to determine the suppressant concentration for extinguishment at consistent conditions of conductive heat losses and sample preheating.

A significant challenge with applying the REED method is determining the actual heat addition rate from the electrical source in a typical energy augmented combustion (EAC) fire in a telecom or data processing fire scenario, so that the appropriate heat flux can be used for comparison in the REED test. Heat addition from radiant heaters is equivalent to that from an electrical short; however, there has been little work done to characterize the heat flows to burning polymers from failing electrical components. Estimates of the heat fluxes in some of the other test methods proposed for Class C fire suppression by clean agents are discussed in section 2.4.2 above and 2.4.3 below, where a comparison is also made of the heat added through the radiant source, or through the wrapped Nichrome wire, and their effect on suppression concentrations.

Other Miscellaneous Test Approaches

Heated Metal Surface Ignition of Premixed Gases

In a pair of papers [12,13], NIST researchers examined the autoignition temperature of premixed hydrocarbon-air mixtures in the presence of various fire suppressants. For ethylene as the fuel, the agents tested were N_2, IG-542[3], HFC-23, HFC-227ea, FC-218, and FC-3-1-10. In addition, methane, ethylene, and propane were tested with CF_3Br, CF_3I, N_2, HFC-227ea, and C_2HF_5. The heated metal surface was nickel foil, which was heated to the range of 760 °C to 1100 °C until auto-ignition occurred. The effect of the suppressant on the autoignition temperature was determined. Also, the concentration of agent required to suppress all hot-surface ignition in the tests with ethylene was determined [13].

For all of the fuels, CF_3Br raised the autoignition temperature by 100 °C to 200 °C at a volume fraction of only 2 %, and the effect for CF_3I was similar for CH_4 and C_2H_4 fuels. HFC-227ea and HFC-125 raised the autoignition temperature for ethylene, but lowered it for methane, and HFC-125 also lowered it for propane. To completely suppress hot-surface ignition of premixed, stoichiometric ethylene-air mixtures, agent concentrations near to the propane-air +10 % inerting concentrations were required for all agents (except in the case of FC3-1-10, which required twice the inerting concentrations). The inerting concentrations are much higher than the suppression concentrations for flames.

These results demonstrate the tendency of both chemically reacting and inert fire suppressants to become less effective at higher temperatures. For example, the inert agents tested in ref [13] required 1.5 to 1.8 times as much agent to suppress the hot surface autoignition as to suppress heptane cup burner flames, whereas the HFCs required 1.9 to 2.3 times as much, and the FCs, 3.1 to 4.8 times as much. While it should be noted that autoignition chemistry is somewhat different from propagating flame chemistry, the lowered effectiveness at higher temperature has also been noted for *flame suppression* in cases of enriched oxygen combustion (as described above in section 2.2.2).

_____

[3] composed of 0.52 N_2, 0.40 Ar, and 0.08 CO_2 volume fractions.

The question naturally arises as to whether the configuration in the tests of Hamins [12] and Braun [13] is realistic with respect to suppression of electrically-energized fire suppression. While overheated metal components of that temperature are possible, as has been noted [7], it does not seem likely that such high temperatures (on the order of 1000 °C) would exist for long. Copper (the likely conductor) melts at 1085 °C; and to maintain the metal at a high temperature without overheating and failing, the power input rate (i.e., current and voltage) would have to be matched very closely with the heat loss rate to achieve a near steady-state. Flammable decomposition products could be present from pyrolyzing polymers, and they could premix with air, in stoichiometric proportions, and impinge on a hot surface; however, a more likely scenario is a diffusion flame.

The most significant results of the hot-surface tests are the quantities of agent required to completely suppress ignition. The primary role of hot metal surface in these tests, besides promoting the autoignition, is to preheat the reactants. As described above in the section Effects of Heat Addition on Suppression (in 3.2.2), heated reactants require more suppressant for extinguishment. The results of Hamins [12] and Braun [13] illustrate that the same appears to be true for autoignition.

NASA WSTF Tests for EVA Suit Wire-Failure Ignition

Recently, NASA technicians examining an Extra-terrestrial Vehicle Activity (EVA) suit which had just been returned from space, found frayed wires [59]. The wires could have been an ignition source during an EVA, with severe consequences in the oxygen-enriched environment of the suit. In order to understand the shorted-wire ignition of materials on the interior of the EVA suits, researchers at the NASA White Sands Test Facility (WSTF) developed three new tests and evaluated them. Two tests, the Multiple Locations Intermittent Arcing Method (Scratch Test) and the Single Location Intermittent Arcing Method (Poke Test) used a needle-like anode electrode to scratch or poke through a test material (fabric) against the cathode. The third method (the Wire-break Test) pressed a thin wire of diameter 0.16 mm to 0.0158 mm (34 AWG to 54 AWG) against the test fabric, and the current (regulated) was increased until wire failure. The power supply for the tests was designed to simulate the voltages and currents of the EVA battery pack, and delivered voltages between 2 V and 35 V, and currents, 0.6 A and 6 A. The third test was found to be the most challenging (and thus most conservative) and was used for the understanding the ignition risk in the EVA suits.

Several findings of the report are of particular value in the present work. Current was generally more important than voltage. Material configuration affected the ignitibility, and frayed materials ignited at much lower power. In the Scratch Test and Poke Test, it was difficult to insure that the arcing event was in intimate contact with the test material (if it were not, ignition did not occur). Similarly, in the Wire-break Test, if the wire were in direct contact with the material, the power required for ignition was much lower than if it were not in contact. A significant finding of the testing was that while all three tests methods could ignite the fabrics, the third test ignited them with the lowest power. The reason was that the wire heating test preheated the surface of the polymer, making fuel species available in the gas phase for ignition. Further, the energy added during the preheating of the wire up to the failure point was three orders of magnitude greater than the energy released during the wire-failure event.

The findings from the NASA WSTF tests are of significance to the electrically-energized fire suppression test desired in the present work. Preheating of the test material prior to its burning and suppression must be considered both in the analysis of the equipment failure mode, and in the development of the test method itself. This finding is further supported by the NASA Glenn Research Center results on suppression of flames over PMMA discussed below.

44

Goldmeer et al. [44] studied the suppression of flames over horizontal cylinders of PMMA in cross-flow, in normal and microgravity. The suppression was achieved through depressurization of the test volume, inducing a flow which blew-off the flames. The sensitivity of the flames to extinguishment was strongly dependent upon the degree of preheating of the PMMA, as well as on the forced convective air flow velocity in the test chamber.

Ruff et al. [45] also studied the extinguishment of PMMA cylinders in crossflow. In their tests, the PMMA cylinder was preheated with a resistive cartridge heater in the center of the PMMA, and CO_2 was added to the air stream to extinguish the flame. The CO_2 extinguishment of PMMA was again found to be sensitive to degree of resistive preheating.

Takahashi and Katta [43] performed experiments and detailed numerical modeling on the suppression of cylindrical polymer samples (PMMA, high density polyethylene, and polyoxymethylene) and cylindrical porous surfaces fueled by methane. The configurations tested are shown in Figure 13 (note that the polymer flames were only tested in the end-up configuration, as in the standard cup burner). They noted that the requirement for heat feedback to the surface for fuel generation resulted in a much different stabilization behavior for the polymer flames, and that with CO_2 added at near extinguishment concentrations, the heat release (i.e., evaluated via the flame size [68]) was much lower. They also noted the importance of melting and dripping.

VTT Electrical Ignition Source Studies

Keski-Rahkonen and Mangs [50] described multi-faceted work to understand risks from electrical ignitions in Finnish nuclear power plants. They performed statistical analysis of fire event data from both nuclear and non-nuclear power plants. For the most common, simple mechanisms of ignition, they performed analytical modeling of the processes to understand the controlling parameters, and also performed experiments of the idealized systems which were modeled. The statistical database provided a useful overview of the problem; however, they found much uncertainty in the data. The fire incidents are reported by the fire officer in command at the fire, not through more comprehensive forensic analysis. Many of the causes are listed as "possible" or "supposed," and a significant fraction of the total are "unknown electrical." Using the statistical fire-event databases, the authors could tabulate the data in terms of the failure mechanism (e.g., overheating, short, ground fault, arcing, etc.) or the failed component (cable, switch, breaker, etc.); however, there was not enough detail to understand the sequence of events leading to the fire, or even the physics of the failure mechanism which finally occurred. This lack of detail necessary to provide the precursor events or to identify the true root causes has been pointed out by others (Madden reference in ref. [50]) with regard to the Sandia and EPRI databases.

The physical modeling and supporting experiments of Mangs and Keski-Rahkonen are useful for understanding electrical ignitions. The authors identified a particular ignition scenario, did some literature review to understand the current state of understanding of the physics, and then made an analytical model of the phenomenon. Finally, they performed supporting experiments. One scenario approached in this way was a loose contact. It was modeled as a plane, cross-sectional source of energy in a wire, which transiently heated the wire (that was subject to convective and radiative cooling). They estimated that for a copper wire of cross-section 0.5 mm^2 to 4 mm^2, only 1 W to 12 W of electrical power was necessary to heat the wire to a temperature which would ignite combustibles nearby (200 K temperature rise), in a time of 5 s to 160 s. They also modeled, and performed experiments on, the heating of an overloaded cylindrical cable, and found that fairly high currents were required to produce

200 K of temperature rise. Based on their calculations, both of these tests, the loose contact and the overheated wire, were deemed to be plausible ignition paths. In some of their supporting experiments, they found as did others [5], that unrestrained PVC insulation on cables in a furnace quickly melted off the cables and dripped away, and hence could not be auto-ignited.

Experiments and modeling were performed for electrical arcs between metal rods. The authors, as did McKenna et al. [5] experimentally found that they had difficulty producing stable arcs, and that the copper quickly melted and failed. They also found that their arcs were so violent that they often blew-off any flames which were formed. They used batteries, however, rather than the DC arc-welding equipment which others have found to be more controllable [60]. Nonetheless, they investigated the physics of electrical arc, and estimated that for copper conductors, 36 V is the minimum for stable arcing, and that higher voltages (e.g., 50 V DC) would produce more stable arcs. (This result is not consistent with the much lower voltages supporting a stable arc in the work of McKenna et al. [4,5] and Khan [52,53,60]. In these latter cases, the presence of decomposing polymeric surfaces and the attendant impurities in the gas-phase arc likely affect the required voltages.)

Finally, the authors overloaded components on printed-circuit boards, and found that only power transistors were likely to lead to ignition of the boards, which could be modeled as piloted ignition.

The data collection, analyses, and experiments in the work of Mangs and Keski-Rahkonen were well thought out and executed. It seems that there is great potential in their approach, but as they also noted, they did not have the resources to pursue all of the fruitful avenues they uncovered. As pointed out by Babrauskas [67], given the importance of electrical ignition to fires in residential structures, relatively little research has been done to understand the basic physical mechanisms which lead from electrical wire faults to structure ignitions. The same is probably true for the physics of electrically-energized fire suppression: there is little fundamental understanding of what the fire scenarios are, making design of a realistic test method somewhat speculative.

2.4.3 Role of External Energy Flux in EAC Fires

Comparison of External Energy Flux in Various Test Methods

In some of the above discussions of the individual tests, estimates were made of the external heat flux (from electrical, radiant, or adjacent flames). It is useful to gather those estimates here, and discuss them, and this is done in Table 6. For each test, the estimated energy flux to the polymer is provided in the Power Added column (and these estimations are described above in section 2.4.2). This parameter is listed as the range of fluxes existing within one test (due to variation with position on the sample for the FMGlobal Electrical Arc Apparatus, or the Overheated Connection Test; or due to variation in time as the sample burns away from the hot wire in the Vertical Polymer Slabs Ignited by a Loop of Nichrome Wire test). For the Overheated Connection Test, the average heat flux on the surface of the rod (45 kW/m^2) is also estimated. For the Printed Wiring Board Test, the average range of 46 kW/m^2 to 80 kW/m^2 is estimated based on the range of voltages for which the stable arcs could be established. The last two columns describe whether the energy added from the external source was added at the burning surface, and whether the heat from the attached flame added heat where the mass loss was occurring. The tests are grouped using the same scheme as in section 2.4.2 above, in the order previously presented.

The tests in which the external heat flux was specified and controlled (middle rows of

Table 6) are the most straightforward to discuss. The geometry for heat addition is more-or-less constant in time and over the burning sample for all of these tests. Hence, there is only a single value of the flux for each test. This is not completely true, since the shape of the burning sample does change in the REED experiment and in the Nichrome wire-wrapped polymer experiments. This variation should ultimately be estimated, but this estimate is beyond the scope of the present project. Shape variations which affect the external heat flux to sample could be important, and ultimately would have to be accounted for or eliminated. Also, as discussed above, there is some uncertainty in the heat flux (due to varying absorptivity to IR and wire contact effectiveness), and these would have to be accounted for more accurately than done here. Nonetheless, the effects of shape variation are expected to be secondary, and can be controlled.

The wire-wrapped polymers had heat fluxes which varied from about 7 kW/m^2 to 24 kW/m^2; presumably, these could be varied from zero to higher values as well. The REED experiment had heat flux of 0 kW/m^2 to 60 kW/m^2, and this range could be extended up to about 100 kW/m^2. The FM Global 50kW Apparatus used 50 kW/m^2, which again, could be extended. In all of these tests, the added energy from the external source was added to the surface, which was the same location where the attached flame added energy to the polymer.

External power input fluxes were also estimated for the Tests Simulating The Failure Mechanism with Suppression, and these are listed in the top set of rows in

Table 6. Several of the tests (FM Global Electrical Arc, Overheated Connection, and Vertical Polymer with Wire) are estimated to produce fluxes which vary over the surface of the tested materials. To estimate the flux from the electrical arcs in the FM Global test, the power input to the arc (1 kW) is dissipated in the (reported) 6.35 mm diameter sphere, and this flux drops off as $1/r^2$ (assumed distances of the polymer to the arc are 2 mm to 25 mm, and these can occur in a single test). This yields a very high flux of 100 kW/m^2 to 3000 kW/m^2, depending upon where on the adjacent burning polymer surface one is considering. For all the other tests in this group, the details of the flux estimates are provided above in section 2.4.2. The flux in the Overheated Connection Test is estimated to vary from about 10 kW/m^2 to 100kW/m^2, depending upon the location on the wire, and the average value is about 45 kW/m^2. These values could have been varied by changing the ring heater temperature; however, a varying heat flux at different positions is a characteristic of the test method. In all cases, the external heat is added to the backside of the polymer; whereas the flame heat is added at the burning surface. Depending upon the time of the test, and the thickness of the insulation, the polymer may be behaving as a thermally thin material, in which case the back-side heat addition is fine. Otherwise, this could affect the influence of the added heat. For the Vertical Polymer Slabs Ignited by Nichrome Wire Test, the energy flux varies because the burning polymer regresses in time from the adjacent hot Nichrome wire. There is probably spatial variation as well (since all parts of the polymer are not equidistant from the wire, and the flame adds heat mostly to different locations than the wire). The external energy is added at the surface of the polymer, but this may not be the same surface at which the flame adds heat. The heat flux from the wire is estimated to range from 6 kW/m^2 to 100 kW/m^2, decreasing in time as the test proceeds. The average heat flux in the Printed Wiring Board Test is estimated to range from 40 kW/m^2 to 80 kW/m^2, varying only according to the voltage and current used to establish the arc. The flux is probably relatively constant during the test. It was not possible to estimate the external heat flux in the Wire Cable Bundle Tests since the power level dissipated in the wire bundle was not given.

Table 6 – Estimated (or measured) heat flux to the burning polymer in various test methods.

Test Method	Power Added (kW/m²)		Heat Added at Burning Surface	
	Range	Average	Auxiliary	Flame
Tests Simulating the Failure Mechanism with Suppression				
FM Global Electrical Arc Apparatus	100 to 3000	na	Y	Y
Ohmic Heating Test		*, dnd	N	N
Overheated Connection Test	10 to 100	*, 45	mostly	Y
Printed Wiring Board Test	na	46 to 80	Y	Y
Vertical Polymer Slabs Ignited by Loop of Nichrome Wire	* 6 to 100	Na	Mostly	N, mod
Wire Cable Bundles with Heated Nichrome Wire	*, dnd	*, dnd	N	N
Test Methods Based on Controlling the External Heat Flux				
Resistively Heated Polymer Samples, Case 1	na	*, 6.8	Y	Y
Resistively Heated Polymer Samples, Case 2	na	*, 16 ± 6	Y	Y
Resistively Heated Polymer Samples, Case 3	na	*, 24 ± 8	Y	Y
Radiantly Enhanced Extinguishing Device (REED)	na	0 to 60	Y	Y
FM Global 50kW Apparatus	na	*, 50	Y	Y
Miscellaneous Test Approaches				
Heated Metal Surface Ignition of Premixed Gases	dnd	dnd	na	na
NASA WSTF Tests for EVA Suit Wire-Failure Ignition	dnd	dnd	Y	na
UL-2127 and UL-2166	0 to 3.2	na	Y	Y
NEBS Fire-Spread Test, Methane Igniter Burner	12 to 32	na	Y	Y

Notes:
  * could be varied but was not
  na : not applicable
  mod : could be modified to do this
  cnd : could not determine
  dnd : did not determine (but could be done with more data from the test)

In the Miscellaneous Test Approaches, the external heat flux for the Heated Metal Surface Ignition tests and the EVA Suit Ignition tests were not determined, but this could have been done. The heat flux in the UL-2127 and UL-2166 tests is estimated (below) through comparison of the MEC in that test with those in the REED. The values are 0 kW/m² (no adjacent flames on some surfaces) to 2.9 kW/m² ± 0.7 kW/m². (where the uncertainty represents the range of fluxes experienced by the single surface). The heat flux from a flame to a burning PMMA polymer surface has also been estimated based on the mass loss rate. The net heat flux on a single vertical piece of free-burning PMMA (10 cm x 10 cm) was found to range from 12 kW/m² to 32 kW/m², depending upon the location on the PMMA, with an average value of 18.2 kW/m² [69], while Tewarson and Pion [16] give an average value of 17.1 kW/m². The heat flux in the NEBS flame spread tests is also shown in Table 6, and the rational is described below.

Another existing test method which can serve as a basis for assessing the heat flux to polymers in electronic equipment is the NEBS Flame Spread Test. As described above, one technical expert suggested that the power level of the ignition fire in the NEBS rack-level test was an appropriate level of power to consider. The logic is as follows. In the NEBS tests, using this ignition source as the initiator, and insuring that the fire-retardant capability of the adjacent components is sufficient to stop propagation, NEBS equipment has demonstrated superior fire resistance. Hence, for the same energy input, if protection is to be obtained by clean agents (instead of by meeting a NEBS standard), then they should be able to suppress materials fires subjected to the same initiating heat flux. The design fire for the NEBS tests is shown in Figure 20. The average power input in the NEBS test is 2.5 kW, over a 330 s time period, with a peak value of 5 kW. The heat flux (power per area) is estimated through analogy with the vertical PMMA slab heat fluxes used above. Since methane-air flames and PMMA-air flames have similar temperature, and their scale and configuration is also similar, the heat flux from the two flame types to a surface is probably similar. Hence in Table 6, the NEBS test is estimated to provide external heat fluxes from the line burner methane flame of 12 kW/m^2 to 32 kW/m^2.

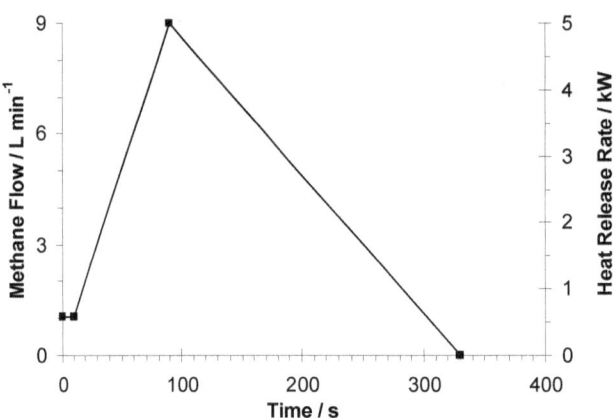

Figure 20 –Input power vs. time for NEBS fire test [70]

As Table 6 shows, in the tests for which it was possible to estimate the energy flux, most fall in the range of 0 kW/m^2 to 100 kW/m^2. This is also the range in which the test results can be compared to those done in the REED device [57]. Hence, it is possible not only to compare the fluxes to which the polymers are exposed, but also the concentration of agent which extinguished the fires for the different tests when subjected to equivalent levels of added heat. To do this, of course, requires test data for the same materials. Unfortunately, many of the tests simulating the failure mechanism of components used different materials from those used in the REED tests. Also, it was not possible to estimate the heat flux in several of the tests, or the extinguishing concentration was given only as a pass/fail results, rather than in terms of a minimum extinguishing concentration. Nonetheless, for the tests which can be compared, the results are given below.

Comparison of Suppressant Requirements in REED and UL Tests of NFPA 2001

Since the existing tests UL-2127 and UL-2166 are the basis of the current NFPA 2001 standard for clean agent concentration requirements for Class A fires (and, hence, Class C fires), it is of interest to determine the effective heat flux to the surface of the PMMA in those tests. Since the UL tests have adjacent vertical PMMA (and other polymer) sheets, there is heat feedback from one burning surface to the other. This heat feedback will increase the suppressant requirement relative to that for sheets in isolation. Taking the suppressant concentrations in the UL tests for comparison, the REED experiments can be examined to determine the heat flux for which that concentration of agent was required. This was done and the results are listed in Table 7. As indicated, the heat fluxes are in the range of (2.2 to 3.6) kW/m^2. That is, the flame on the first sheet of PMMA appears to impose a heat flux of (2.2 to 3.6) kW/m^2 on the second sheet. Hence, REED heat fluxes of the order of (2.2 to 3.6) kW/m^2 replicate the amount of agent for suppression of the flames from the adjacent PMMA sheet in the UL-2127 and UL-2166 Class A test.

Table 7 – NFPA2001 Class A PMMA suppressant requirement [71], and the REED heat flux [11] at that concentration, for several agents.

Agent	NFPA 2001 Class A	REED	
	X_{ext}	X_{ext}	Flux
HFC-227ea	0.058	0.058	2.2 kW/m^2
HFC-23	0.11	0.11	3.1
IG-541	0.326	0.326	3.6

Comparison of Suppressant Requirements in REED and Wire-Wrapped PMMA Tests

The agent required to extinguish PMMA samples with heat added by wrapping with Nichrome wire [57] can be compared to that with heat added radiantly (i.e., the REED experiment [11]). As a rough estimate, we can assume for the wire-wrapped PMMA that 50% of the energy dissipated in the Nichrome wire goes into the polymer (i.e., half of the heat flow inward, half outward). This would be a lower limit, while an upper limit would be to assume that 100 % of the energy to the wire makes it into the PMMA. These estimates provide the energy flux into the polymer, and the extinguishing concentration of agent is given [57] for these conditions. Using these values of the heat flux from the wire-wrapped PMMA experiments, the REED extinguishing conditions at those values of the heat flux are found from the data of reference [11]. The equivalent heat flux values are 6.8 kW/m^2 [9], 11 kW/m^2 to 22 kW/m^2 [9], and 16 kW/m^2 to 32 k/m^2 [57], and the agents considered are HFC-23, HFC-227ea, FC-2-1-8, and FC-3-1-10. The extinguishing volume fractions of the agents at the equivalent values of the heat flux are shown in Figure 21; Figure 22 shows the data of Figure 21 together with the data for the agent IG-541. In the figures, the average value of the NIST and 3M data (when available) are given for the REED device [57], and the error bars for the Y-axis represent the limits of these two results. For the wire-wrapped PMMA, the values of the extinction concentration are given for flames with some re-flash, or complete inertion, and the error bars in the X-axis represent the limits of these values. As the figures show, the two measurements (with the error bars shown) agree with each other within about 15 %.

Adding energy to the polymer surface with the REED device appears to be equivalent to adding by wrapping with Nichrome wire. That is, as shown above, the wire-wrapped PMMA extinguishment results give extinguishing concentrations in close agreement with the REED device. In addition, the NFPA 2001, UL-2166 and UL-2127 tests give results for multiple PMMA sheets which are also consistent with the REED test results (at equivalent levels of imposed heat flux).

The agreement in Figure 21 and Figure 22, while quite good, could be improved even more if other factors were controlled. For example, the configurations and air flow (and hence flame stabilization) differ somewhat between the test methods, and these can affect the suppression process. Also, as described above in section 2.2.1, the energy flux of interest is really the net flux to the polymer, not the imposed external flux. Nonetheless, these additional effects are expected to be of secondary importance for the tests compared here, but in other configurations, the additional effects may be more significant.

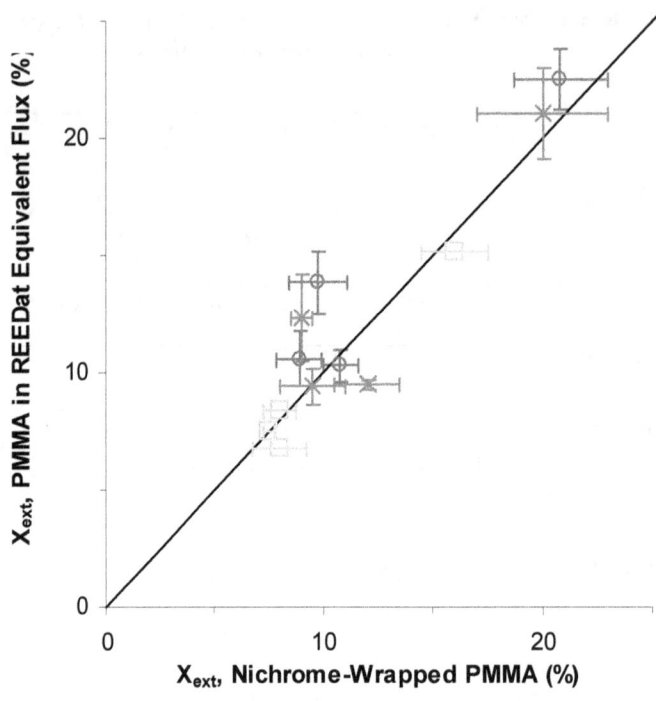

Figure 21 – Agent (HFC-23, HFC-227ea, FC-2-1-8, and FC-3-1-10) extinguishing volume fraction for PMMA with Nichrome wire heating, or at an equivalent heat flux achieved in the REED device (○: ref [57]; □ and +: 48 W and 192 W cases of ref. [9]).

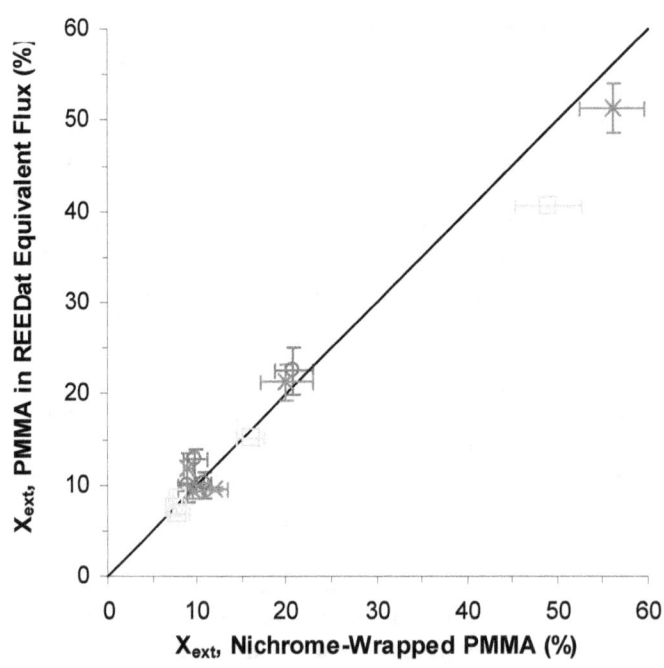

Figure 22 - Agent extinguishing volume fraction with the agents in Figure 21 as well as with IG-541 for PMMA with Nichrome wire heating, or at an equivalent heat flux achieved in the REED device (○: ref [57]; □ and +: 48 W and 192 W cases of ref. [9]).

Fundamentally, if energy were added to the surface of a burning polymer, it will increase the suppressant requirement for extinguishment. It would not matter too much if the energy were added radiantly, with conducting hot surfaces, or from adjacent flames. The primary challenge is to understand the conditions of actual energy-augmented combustion electrical fires sufficiently well to determine the appropriate energy flux levels (in the REED or any other test method).

Comparison of Suppressant Requirements in Other Tests

Two additional tests can be compared with the REED results: the FM Global Fire Propagation Apparatus tests and the Vertical Polymer Slabs Ignited by a Loop of Nichrome wire test, both of which were performed with PMMA. The FM Global FPA tests and the REED tests are shown in the upper curves of Figure 23 (for N_2 extinguishment). As indicated, the results agree well. For the vertical PMMA with the hot wire, the extinguishing volume fraction of HFC-227ea was given as < 0.058. It is difficult to estimate the heat flux for this condition, since the separation between the wire and the PMMA at the end of the test (when the suppression occurs) is not known. Nonetheless, using the REED results for this agent with PMMA [58] indicates that a volume fraction of HFC-227ea of <0.058 implies a heat flux of < 7.8 kW/m^2. Using Figure 17, we see that this corresponds to a separation distance of >2.9 mm. While the exact details of the conditions is this test are hard to specify (as described in section 2.4.2), the two tests give consistent results. It should also be kept in mind that the phenomena described above (stabilization and heat losses) are not completely controlled between the tests, so exact agreement is not expected.

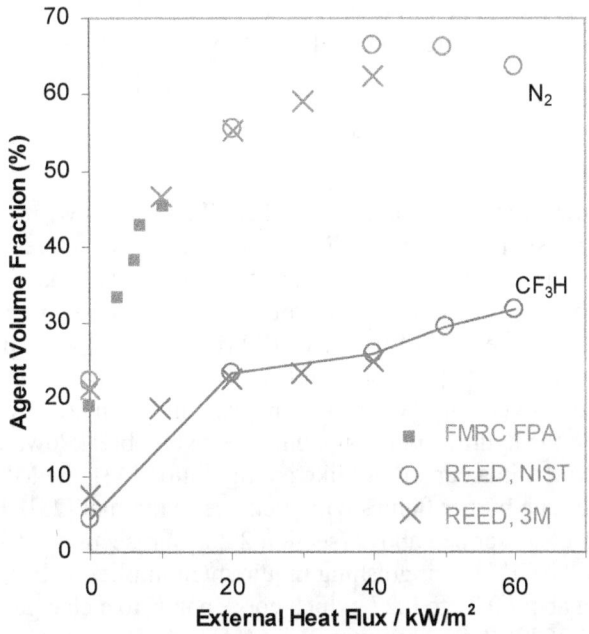

Figure 23 – HFC-23 or N_2 volume fraction for extinguishment of PMMA in the REED device as a function of external heat flux (NIST and 3M results are given). Tests with N_2 extinguishing PMMA in the FM Global FPA [16] are also shown.

53

Effect of External Added Energy on Suppression Concentration

General Comments

It would have been very interesting to have results, from all of the test approaches described above, at variable amounts of external heat input. This would be useful because: 1.) The amount of external heat added in real, failing electrical components is not well known, so understanding the sensitivity of the suppressant requirements to this parameter is important, and 2.) Data from all the tests at the same values of the external heat flux would allow cross-comparison of the test results, which probably could be unified with the relevant fundamental parameters. Of course, these test results would be needed for the same materials. (Preheating and stabilization conditions would have to be accounted from as well, but to first approximation, external heat flux and material type are probably most important).

The influence of heat addition on the suppressant concentration is most clearly shown by the REED test results, since these were done for a range of external heat flux values. For all agents tested, the required agent volume fraction for suppression increased rapidly at low external heat flux. Such results are illustrated for N_2 and CF_3H in Figure 23. As the tests for other agents show [58], doubling the suppressant concentration in the REED device requires low heat fluxes, around 9 kW/m^2 for inert agents, and around 6.4 kW/m^2 to 8.1 kW/m^2 for HFCs. At higher external heat flux, the amount of agent tended to reach a maximum, beyond which the increase in agent requirement was little or none. This is consistent with the findings in the suppression tests of arc-heated PVC cables by CF_3Br [60], in which the flames were extinguished even at very high external heat fluxes.

Influence of Gas-Phase vs. Solid-Phase Heating

As indicated above for PMMA (Figure 23), an external heat flux of 6.5 kW/m^2 is required to double the amount of CF_3H necessary for suppression in the REED device. It is of interest to estimate the external heat flux which would be required if the heat was added only to the gas-phase products rather than to the polymer surface. This can be done as follows. As a rough estimate, one can assume that the effect of the gas-phase temperature on the suppression of flames of PMMA combustion is similar to that of methane combustion. This is a reasonable assumption since it has been demonstrated that flame suppression behavior of hydrocarbon fuels is very similar because they are all dominated by the same chain-branching radical reactions, common to the hydrocarbon fragments created by breakdown of the initial fuel molecule [72]. Also, for PMMA in a cup burner like configuration [43], the MEC for CO_2 was found to be essentially the same as for cup burner flames with methane as the fuel [25]. Hence, using the experimental cup burner results described above (section 2.2.2, sub-section: Effects of Heat Addition on Suppression), a doubling of the CF_3H extinguishing requirement implies a change in the oxygen volume fraction in the oxidizer from about 0.21 to 0.27, which corresponds to a change in the final temperature from 2230 K to 2456 K, or 226 K. A free-burning PMMA sample 10 cm x 10 cm x 2.54 cm, horizontal or vertical, loses mass at about 8 g/m^2/s [73], and its complete combustion requires 2.5 mol-air/m^2/s. To raise this amount of air by 226 K requires a heat flux of 17 kW/m^2. Hence, a local energy density of 17 kW/m^2, added to the gas phase, would likely raise the final temperature of the products by about 226 K and cause an increase in the amount of CF_3H for extinguishment by a factor of two. This is two or three times the external heat flux that was required in the REED device (which added the heat to the polymer surface). Hence, adding energy to the polymer surface appears to have a much larger effect (by about two or three times) than adding the energy to the gas phase. This is likely due to the positive feedback of the system: heat added to the polymer surface creates more fuel, which creates a larger flame, which increases the heat feedback, etc.

One question which has been raised is whether the relevant parameter to duplicate in a test method is the failed component temperature or the heat flux from the failed component. If the failed component is being considered as a possible ignition source, then temperature is important. However, in the context of the current project, it can be assumed that ignition has already occurred; i.e., a fire has been detected, a clean agent is about to be deployed, and the question is how much agent needs to be added; i.e., does the failed equipment add energy to the system and does this energy affect the quantity of agent required for suppression. Hence, considering the failed component not primarily as an ignition source, but as a source of additional energy, is more relevant.

From the discussions above in section 2.2.1, the main effects of added heat are to raise the temperature of the burning polymer (or the gas phase). The final temperature of the polymer (or the gas phase) is determined by an energy balance, and that is controlled by the heat flows. Of course, *one* parameter affecting the heat flow is the temperature. In general [74], the amount of heat transferred is proportional to the temperature difference, the area, and the heat transfer coefficient $\dot{q} = h\ A\ (T - T_0)$; radiation transport is an exception in which the heat transfer is proportional to T^4. The only limitation on the temperature of the heat source is that it be higher than the object into which the heat is flowing. Since polymers typically decompose at relatively low temperatures (300 °C to 500 °C), and heat can be added to air at the air inlet temperature, or to the polymer at lower temperatures (preheating), the heat source temperatures need not be excessively high. Of course, high heat fluxes (power per unit area) typically require high temperature sources, but it is the heat flux which is the relevant parameter. To some extent, that is influenced by the power available in the equipment.

2.4.4 Recommended Test Method

Overview

Based on the energy fluxes from the simulated equipment fires, and the thoughts of the technical experts, significant energy addition to the burning materials can occur for some configurations in the field (either from adjacent flames, large-area initial ignition, conductive heat transfer from a component, or radiant heat transfer from failed components). Hence, a test procedure must include this parameter as a variable. A review of the fundamental aspects of suppressed flames over burning polymers with energy addition indicates that the forms of the added heat, radiant or conductive (e.g., from a resistive source) are equivalent. It was not possible in the present project to gather enough information to understand the appropriate energy flux level from energy-augmented combustion for all electrically energized equipment fires (nor did the review of the literature indicate that anyone else has done this, although progress has been made). Nonetheless, the review suggests that a test procedure based on radiantly heated polymer samples with two limiting fluxes is appropriate, as discussed below.

Considering the background material outlined in the present report, it is possible to specify the desired qualities in a test procedure for determining the suppressant concentrations necessary for extinguishing a wide range of electrically energized equipment fires, and these are listed in Table 8.

Table 8 – Desired properties in test method to determine suppressant quantities for Class C fires.

1. Not too expensive or time consuming.
2. Accurate and repeatable.
3. Applicable to a wide range of electrically energized equipment fires.

4. Known concentration of agent at the flame stabilization point when extinguishment occurs
5. Well-characterized, and strong flame stabilization (i.e., consistent and tractable configuration, known and consistent flowfield and gas velocity).
6. Understood, consistent, and repeatable level of material preheating and conductive heat losses.
7. Well-characterized and variable (down to zero) amount of added energy to the burning polymer surface.
8. Desired level of radiant input from simulated adjacent flames can be incorporated.
9. A range of materials can be used (especially realistic ones), that can be formed to the shape required by the test method, accounting for the desired level of melt-drip containment.
10. Independent ignition source, which adds little additional energy to the combustion process, with controllable start and stop times.

If one wanted a relatively conservative test method, it would have:

1. *Constant (low-power) ignition source.* In the application to be protected, a recurring ignition source could exist, separate from the failed component adding the energy. These could include a small arc discharge from a shorting component, an attached flame, or an adjacent flame. Hence, a continuous ignition source should be included.

2. *Optimum ventilation and good flame stabilization.* Since both the ventilation condition and stabilization are very configuration dependent, and such a wide variation in shapes and configurations is possible in the field, one must consider that both good flame stabilization and adequate air flow could be present.

3. *Heat input from and adjacent flame.* The arguments in item 2 above apply here. Unless one is considering just a particular piece of equipment, the possibility exists for adjacent burning materials.

4. *Realistic materials, thicknesses, and configurations.* While test methods can be developed and run with any materials (for example PMMA, which is nearly a standard for materials flammability studies), the arguments about the volume fraction of agent required for suppression should be based the materials contained in the equipment to be protected.

5. *Variable external heat flux.* Since there are expected to be a wide range of possible failure modes in electrically energized equipment fires, it is important to determine the sensitivity of the suppressant requirements to the flux of added energy.

The properties in Table 8 constrain the test method. For example, items 4 and 5 imply that the device has a chimney, with controllable, quasi-steady agent concentrations and gas velocities. Items 6 and 9 imply that the sample might not be too small, or that special attention to transient heat losses will be required. Items 7 and 8 concern the energy addition, and require discussion. A radiant source (as in [11,52,53]) is possible, as is a Nichrome wire-wrapped sample. In the latter case, intimate contact could be insured with pre-loading of the wires against the polymer; however, if the material intumesces, or melts excessively, maintaining the proper location of the wires would be difficult. Also, as discussed above, the amount of energy which makes it into the polymer is not straightforward to estimate a priori, although measurements could be made (e.g., mass loss) to characterize this heat transfer. One could use a thermally-thin sample around a resistive cartridge heater, but sample preparation for a range of material types might become challenging. Of these possibilities, radiant energy input appears to be the easiest to apply since it can be used with a wide range of material types, and excessive custom sample preparation is not required. Item 10 on the list implies that a small pilot flame or low-energy spark igniter with programmable duration is available.

One can't completely specify the test method until the supporting characterization experiments have been done. For example, we do not yet know the appropriate input power levels. Nonetheless, based on the energy fluxes estimated for the test procedures proposed to date to describe electrically energized equipment fires (see sections 2.4.2 or 2.4.3), we can propose a test method, and provide two realistic energy fluxes as upper and lower limits.

Test Method Configuration

The basic configuration recommended is a horizontal sample, with insulated bottom and edges, in a chimney with a radiant heat source above. The chimney allows controllable oxidizer flow velocity and composition (set by the operator), and the agent concentration is increased slowly until extinguishment occurs. The radiant source allows direct examination of the effect of added energy on the extinguishment. To contain any melted sample, the test material is wrapped on the bottom and sides with aluminum foil. The foil provides the added benefit of maintaining the heat flux to the sample relatively constant as the sample surface regresses. A small methane-air pilot flame 1 cm above the surface and slightly in from the edge provides a continuous ignition source. The sample size and thickness is variable. The thickness of the materials should be that expected to be used in the equipment. The perimeter should be as small as possible to allow a smaller heater and chimney. A size of 10 cm x 10 cm is large enough, but it is worth considering a smaller sample.

A configuration like the FM Global Fire Propagation Apparatus [16] seems reasonable. A variation of that apparatus used for testing of Halon 1301 extinguishment of PMMA with added heat [52] is shown in Figure 24. It has a chimney, controllable agent/air concentration and flow, controllable energy input, and reasonable sample size (10 cm x 10 cm x variable thickness). All of the parts of the device typically used for measuring heat release rate would not be necessary (although in some situations they would be useful for research purposes.) A disadvantage is that the heaters in the FM Global device use a relatively high-temperature source, so the absorptivity of the polymer surface, which could be coated with carbon black initially, might change during the test. Alternatively, a configuration like the controlled-atmosphere cone calorimeter [75] is a possibility; however, the halogens in the clean agents might attack the cone heater too rapidly, and the quantities of agent required might be too large (the nominal air flow is much larger through the cone). The REED device is a possibility if the sample size can be shown to be large enough (or if the sample holder can be modified to control heat losses), or if larger samples are used, and a continuous ignition source is added. The REED device is shown in Figure 25. (A guard flow of nitrogen from the outer chimney in the REED device is intended to reduce corrosion of the cone heater—and this could be done in the cone calorimeter as well. While no adverse effects of acid gases on the heater were reported in the REED tests, the possible need for further remedial action should be kept in mind.)

In this test method with a radiantly heated sample in a flow chimney, the power input would need—at some point in the agent specification process—to be connected to the power levels actually expected in the protected equipment. This concern has been expressed in earlier work [6]. It is believed, however, that determining the appropriate (if any) power input level which occurs in applications in the field is required for any test method. For example, in the Tests Simulating the Failure Mechanism with Suppression described in section 2.4.2 above, engineering judgment was required to determine the appropriate power level to assign in the test method to represent failures in the field. As described above, these tests can serve as a basis for specifying the radiant flux levels in the proposed test. A similar process will be required for applying the proposed test method to specific threats in energized electrical equipment for which laboratory simulations have not yet been performed. Nonetheless, two limiting cases can be specified already, as described below.

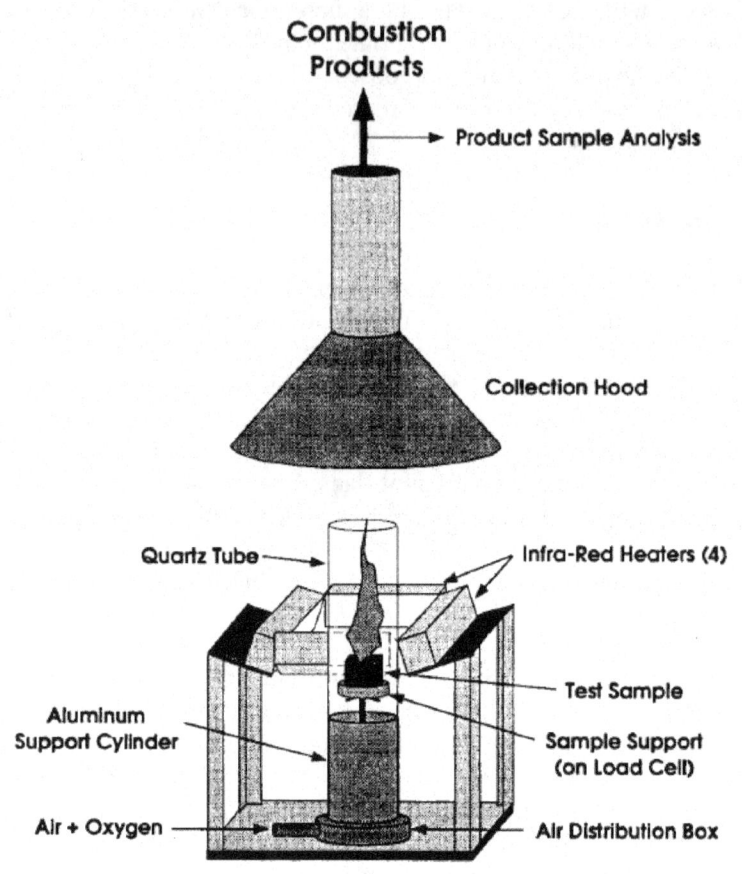

Figure 24 – FM Global flammability apparatus (50 kW scale) [52].

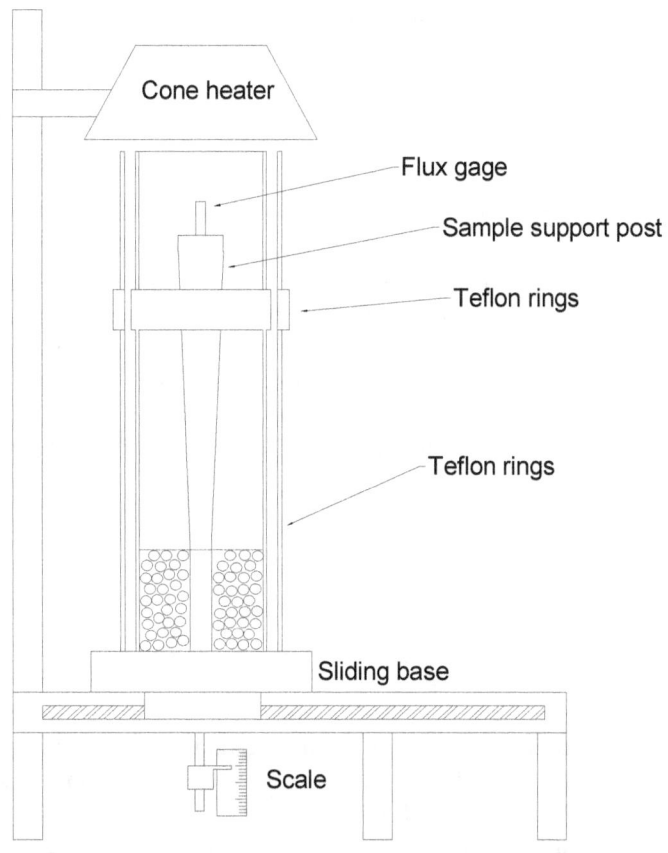

Figure 25 - Radiantly Enhanced Extinguishing Device (REED) [11].

<u>Added Power Levels</u>

Two power levels are recommended initially. The first, 20 kW/m², represents a power input to the burning material only from an adjacent flame on similar burning material. No power input from an electrical source is simulated here, only the possible scenarios in which the burning material is in a configuration where the burning surfaces interact with each other. This is essentially a lower limit of plausible heat flux. The source for this number is the estimate of the heat flux from a flame on PMMA to the burning material (1.2 kW/m² to 3.2 kW/m²) [16,69], described above. This number is likely to be similar for other burning polymers (besides PMMA) since the adiabatic flame temperature of hydrocarbon flames is similar, and the heat transfer is dominated by the flame temperature.

The second recommended heat flux corresponds to well-heated, melted cables, with a continuous electrical energy source. To estimate the flux appropriate for this scenario, we can examine the REED results [58]. In those tests, the required agent for extinguishment increases roughly linearly at low flux, and then asymptotes to a constant concentration at higher flux. The heat flux at which the agent concentration asymptotes (generally around 50 kW/m²) represents the heat flux at which the heat losses (transient heating, edge losses, re-radiation) are no longer affecting the marginal burning rate. That is, below about 50 kW/m² the flame temperature likely increases with flux, whereas above 50 kW/m² the flame temperature is probably a maximum, constant value. Some justification of this number also comes from the estimates of heat flux from simulated failed electronic components to the burning polymers in the section above: Tests Simulating the Failure Mechanism with Suppression. The heat flux values, listed in Table 6, are (10 to 100) kW/m^2 in the overheated connection test, (40 to 80) kW/m^2 in the printed wiring board test, and (6 to 100) kW/m^2 in the Nichrome Wire Near PMMA test. The estimates are listed as ranges since the heat flux varies with position on the polymer. For the Overheated Connection Test, the <u>average</u> heat flux is at least 45 kW/m^2. Based on these estimates, heat flux values of 50 kW/m^2 can be representative.

A few additional details remain in applying the proposed test method to the problem of suppression of fires in energized electrical equipment. In the test procedure, the role of preheating will require attention. For example, at a given input flux, the amount of preheating prior to suppression will affect the MEC. When repeating the suppression tests, it should be possible to quickly identify the approximate MEC, so that consistent values of the preheating time (for example, two minutes), can be used when the flame over the sample is suppressed. In addition, tests must be conducted using realistic materials. Finally, as described above, for cases of added electrical energy between the limit cases, the amount of power possible for a failed component (and the associated energy flux) must be estimated and then specified for the radiant heating tests.

2.5 Recommended Research

The approach intended for the present project at the outset: literature review, survey of industry experience with suppression of electrically energized equipment fires and fire event databases, threat definition, and test method evaluation and development is sound. Nonetheless, the work to date suggests that a key additional component is needed. Because the statistical fire data as well as industry surveys are not expected to provide the level of detail necessary to understand the physical phenomena well enough to specify a realistic test method, expected failure modes must be identified, and then studied through laboratory experiments and analytic modeling. Since the range of equipment environments and failures is too wide to be handled together initially, this must be done on a case-by-case basis until general principles can be developed. The steps recommended to approach this problem are listed in Table 9 below.

Table 9 – Recommended future research to specify a test procedure for suppression of electrically energized equipment fires.

1. Do a survey, as broad and as deep as possible, to understand the fire events which have occurred in the equipment of interest.
2. Collect statistical data on fire events in electrically energized equipment fires. Using the statistical data and survey, categorize the failures in terms of equipment classes (e.g., energized DC power cable fire, energized AC power cable fire, power supply failure, PWB circuit short, battery room fire, etc.)
3. Consult with equipment experts (perhaps at organizations like the National Electrical Equipment Manufacturers, NEMA) who understand the equipment failures well, and can provide insight on what their typical failure mechanisms are.
4. Categorize the fire events in terms of the relevant parameters (using Table 3 as a guide), and develop as many Example Cases as necessary to cover the classes of fire events of interest.
5. For specific Example Cases, model the failure and do laboratory experiments in support of the modeling, to understand the importance and values of parameters in Table 3 for that particular Example Case.
6. Conduct suppression studies in the experiments developed above which mimic the Example Cases.

7. Select an appropriate power level for the radiant test procedure, and refine the standard test method procedure outlined in section 2.4.3 above (or select a new one) to be appropriate for the Example Case examined in the modeling and experimental work.
8. Categorize the Example Cases in terms of the relevant parameters, and group them together, if possible, in similar test procedures.

This is basically a hybrid approach. It uses the statistical database examination of Keski-Rahkonen and Mangs, the survey or McKenna et al., the specific design failure approach of McKenna et al. and Keski-Rahkonen and Mangs, and modeling of Keski-Rahkonen and Mangs. Yet after the failure is understood, is seeks to put the relevant fundamental parameters into a test method in a controllable way, as in the work of Babrauskas, Grosshandler and Donnelly, Smith and Rivers, and Tewarson and Khan. The key feature is that it is approaching the problem initially in terms of specific equipment failures, seeking to understand the controlling parameters in that case, and using these specific example cases to determine the relevant values of the controlling parameter. Then, the appropriate test method—which has the correct cut-off values of the controlling parameters, can be specified. Only after this has been done for a number of specific failure modes (example cases) will it be apparent whether some general principles or approaches to the test methods can be applied, and what the appropriate grouping of the recommended test procedures is for the range of threats expected in the field.

3. Conclusions

The problem of suppression of fires over condensed phase materials with heat addition from an electrical source has been reviewed. Discussions with industry technical experts in fire suppression have been outlined, and a number of cases studies have been presented. Suggested test methods for determining the suppressant requirements for fires in electrically energized equipment have been reviewed. Approximate estimates of the energy fluxes in those tests have been made and compared. The major conclusions of the present study are listed below.

1. The material burning rate (and suppressant concentration) is very sensitive to the heat feedback (especially near extinction); hence, changes to the net external heat flux (from any source) will affect the minimum extinguishing concentration of suppressant.

2. Based on analysis of the test procedures simulating electrical failures proposed to date, the magnitude of the external heat flux in most of the tests is similar in magnitude to that which can be obtained with radiant heating experiments.

3. In cases where the external heat flux could be estimated, the materials were the same, and the flame stabilization was similar, the suppressant concentration measured with the different tests agreed with each other reasonably well.

4. Many of the test methods previously proposed do not combine the relevant parameters in ways which produce the most conservative (yet plausible) test.

5. A test based on an external radiant heat flux source, a large sample (10 cm x 10 cm) in a chimney, and realistic materials is a good starting point for a test procedure.

6. Two radiant flux levels in the proposed test method are suggested:
   a.) an incident flux of 20 kW/m^2 as a lower limit, representing the heat input without any electrical augmentation, but with an adjacent flame on similar burning material (which enhances the burning);

b.) an incident flux of 50 kW/m^2 for cases representing sufficient electrical energy to bring the polymer to its decomposition temperature and maintain it there (in the absence of the flame). An example of this would be an energized cable fire.

7. To assign appropriate energy flux levels for electrical power addition intermediate between these two limits, better understanding of specific electrical failure modes is required.

8. An approach to determine the realistic power levels for situations between the limiting cases is suggested. The first steps are to survey the fire suppression industry and to collate statistical data on electrical fire incidents. These must be followed, however, by three additional steps: obtaining input from electrical equipment hardware experts (or experts in forensic investigation of electrical failures), performing laboratory experiments, and modeling to simulate the likely failure events so that the values of the relevant controlling parameters can be estimated.

9. The appropriate value of the external heat flux can be determined with either a prescriptive- or performance-based approach. As in item 8.) above, Example Cases for specific failure modes in the field can be developed, and then used to specify the heat flux and materials for the test procedure (to protect that threat). Alternatively, the test can be performed for a material at a range of fluxes, and it will be up to the system designer to determine the flux to be experienced in a failure for a specific piece of equipment, and the agent design concentration would then be specified for that application.

4. Acknowledgements

The helpfulness of the Technical Panel and Industry Sponsors who provided input is greatly appreciated; many of the contacts they recommended were extremely helpful, and we are much obliged to them for their input. We appreciate the helpful guidance of Casey Grant throughout the project. We are indebted to Soonil Nam and Robert Kasiski of FM Global for providing the detailed case histories, and follow-up information, and to Richard Gann for a careful review of the manuscript.

5. References

[1]   NFPA 2001 standard on clean agent fire extinguishing systems 2008 edition. Quincy, MA: NFPA; 2007.

[2]   McMenamin, D, "Electrical fires and the power disconnect issue", In: 19th International Telecommunications Energy Conference, INTELEC 97, IEEE, Earth, (1997).

[3]   Robin, ML. Fire protection in telecommunication facilities. Process Safety Progress 2004; 19:107-111.

[4]   McKenna, LA, Gottuk, DT, DiNenno, PJ, "Extinguishment tests of continuously energized class C fires", In: Gann, RG, Whisner, KC, Burgess, SR, Reneke, PA, eds., Papers from 1991-2006 Halon Options Technical Working Conferences (HOTWC), CD-ROM, NIST SP 984-4, HOTWC-1998, National Institute of Standards and Technology, Gaithersburg, MD, (2006).

[5]   McKenna, LA, Gottuk, DT, DiNenno, PJ, Mehta, S. Extinguishment tests of continuously energized Class C fires using HFC-227ea. Baltimore, MD: Hughes Associates, Inc.; 1998.

[6]   Robin, ML, Shaw, B, Stilwell, B, "Development of a standard procedure for the evaluation of the performance of clean agents in the suppression of Class C fires", In: NFPA Suppression and Detection (SUPDET), SUPDET-2007, National Fire Protection Association, Quincy, MA, (2007).

[7]   Robin, ML, Stilwell, B, Shaw, B, "Summary of ongoing Class C fire research for the purpose of identifying and evaluating Class C fire risks and suppression needs in modern data centers, internet service providers and telecommunications facilities", In: NFPA Suppression and Detection (SUPDET), SUPDET-2008, National Fire Protection Association, Quincy, MA, (2008).

[8]   Driscoll, MR, Rivers, PE. Clean extinguishing agents and continuously energized circuits. In: Annual Conference on Fire Research: Book of Abstracts. Gaithersburg, MD: National Institute of Standards and Technology; 1996. 51-52.

[9]   Niemann, R, Bayless, H, Craft, C, "Evaluation of selected NFPA 2001 agents for suppressing Class "C" energized fires", In: Gann, RG, Whisner, KC, Burgess, SR, Reneke, PA, eds., Papers from 1991-2006 Halon Options Technical Working Conferences (HOTWC), CD-ROM, NIST SP 984-4, HOTWC-1996, National Institute of Standards and Technology, Gaithersburg, MD, (2006).

[10] Steckler, KD, Grosshandler, WL, Smith, WL, Rivers, PE. Clean agent performance on fires exposed to an external energy source. In: Beall, KA, editor. Annual Conference on Fire Research: Book of Abstracts. November 2-5, 1998. Gaithersburg, MD: National Institute of Standards and Technology; 1998. 127-128.

[11] Donnelly, MK, Grosshandler, WL. Suppression of fires exposed to an external radiant flux. NIST IR 6827, Gaithersburg MD: National Institute of Standards and Technology; 2001.

[12] Hamins, A, Borthwick, P. Suppression of ignition over a heated metal surface. Combustion and Flame 1998; 112:161-170.

[13] Braun, E, Womeldorf, CA, Grosshandler, WL. Suppression concentration of clean agents exposed to a continuously energized heated metal surface. Fire Safety Journal 1999; 33:141-152.

[14] Smith, DM, Niemann, R, Bengtson, G, "Examination and comparison of existing Halon alternatives and new sustainable clean agent technology in suppressing continuously energized fires", In: Gann, RG, Burgess, SR, Whisner, KC, Reneke, PA, eds., Papers from 1991-2006 Halon Options Technical Working Conferences (HOTWC), CD-ROM, NIST SP 984-4, HOTWC-2001, National Institute of Standards and Technology, Gaithersburg, MD, (2006).

[15] Bengtson, G, Flamm, J, Niemann, R, "Update on the examination and comparison of existing Halon alternatives and new sustainable clean agent technology in suppressing continuously energized fires.", In: Gann, RG, Burgess, SR, Whisner, KC, Reneke, PA, eds., Papers from 1991-2006 Halon Options Technical Working Conferences (HOTWC), CD-ROM, NIST SP 984-4, HOTWC-2002, National Institute of Standards and Technology, Gaithersburg, MD, (2006).

[16] Tewarson, A, Pion, RF. Flammability of Plastics .1. Burning Intensity. Combustion and Flame 1976; 26:85-103.

[17] Tewarson, A. Generation of heat and chemical compounds in fires. In: Beyler, CL, Custer, RLP, Walton, WD, Watts, JMJr, Drysdale, D, Hall, JRJr, Dinenno, PJ, editors. SFPE Handbook of Fire Protection Engineering. Quincy, MA: National Fire Protection Association; 1995. 3-53.

[18] Rhodes, BT, Quintiere, JG. Burning rate and flame heat flux for PMMA in a cone calorimeter. Fire Safety Journal 1996; 26:221-240.

[19] Linteris, G. Numerical simulations of polymer burning rate and the inferred effective heat of gasification. Fire and Materials 2008; in preparation.

[20] Quintiere, JG. Principles of Fire Behavior. Albany, NY: Delmar; 1998.

[21] Williams, FA. A unified view of fire suppression. Journal of Fire and Flammability 1974; 5:54-63.

[22] Linan, A. The Asymptotic Structure of Counterflow Diffusion Flames for Large Activation Energy. Acta Astronautica 1974; 1:1007-1039.

[23] Trees, D, Seshadri, K, Hamins, A. Experimental Studies of Diffusion Flame Extinction With Halogenated and Inert Fire Suppressants. In: Miziolek, AW, Tsang, W, editors. Halon Replacements: Technology and Science. Washington, D.C.: ACS Symposium Series 611, American Chemical Society; 1995. 190-203.

[24] Zebetakis, MG. Flammability characteristics of combustible gases and vapors. Bulletin 627, Washington: U.S. Dept of the Interior, Bureau of Mines; 1965.

[25] Takahashi, F, Linteris, GT, Katta, VR. Extinguishment mechanisms of co-flow diffusion flames in a cup-burner apparatus. Proc Combust Inst 2006; 31:2721-2729.

[26] Takahashi, F, Linteris, GT, Katta, VR. Extinguishment Mechanisms of Microgravity Diffusion Flames in Air and Oxygen-Enriched Streams with Dilution. Combustion Flame 2008; in preparation.

[27] Takahashi, F, Linteris, GT, Katta, VR. Vortex-coupled oscillations of edge diffusion flames in coflowing air with dilution. Proc Combust Inst 2006; 31:1575-1582.

[28] Linteris, GT, Gmurczyk, GW. Prediction of HF formation during suppression. In: R.G.Gann, editor. Fire suppression system performance of alternative agents in aircraft engine and dry bay laboratory simulations. Gaithersburg, MD: National Institute of Standards and Technology; 1995. 201-318.

[29] Linteris, GT, Truett, L. Inhibition of premixed methane-air flames by fluoromethanes. Combust Flame 1996; 105:15-27.

[30] Linteris, GT, Burgess, DR, Babushok, V, Zachariah, M, Tsang, W, Westmoreland, P. Inhibition of premixed methane-air flames by fluoroethanes and fluoropropanes. Combust Flame 1998; 113:164-180.

[31] Zallen, DM, Morehouse, ET, Jr., "Fire extinguishing agents for oxygen-enriched environments", In: Schroll, DW, ed., Third International Symposium on Flammability and Sensitivity of Materials in Oxygen-Enriched Atmospheres, Flammability and Sensitivity of Materials in Oxygen-enriched Atmospheres: Third Volume, American Society for Testing and Materials, Philadelphia, PA, (2008).

[32] Katta, VR, Takahashi, F, Linteris, GT. Fire-suppression characteristics of CF_3H in a cup burner. Combust Flame 2006; 144:645-661.

[33] Takahashi, F, Linteris, GT, Katta, VR, "Extinguishment mechanisms of cup-burner flames", In: AIAA Paper 2008-0745, 44th Aerospace Sciences Meeting and Exhibit, AIAA, Reston, VA, (2006).

[34] Fenimore, CP, Martin, FJ. Flammability of polymers. Combustion and Flame 1966; 10:135-139.

[35] Fenimore, CP, Jones, GW. Modes of inhibiting polymer flammability. Combustion and Flame 1966; 10:295-301.

[36] Fenimore, CP, Martin, FJ. Candle-type test for flammability of polymers. Modern Plastics 1966; 12:141-192.

[37] Bajpai, SN. An Investigation of the Extinction of Diffusion Flames by Halons. Journal of Fire and Flammability 1974; 5:255-267.

[38] Hirst, B, Booth, K. Measurement of flame extinguishing concentrations. Fire Technology 1977; 13:296-315.

[39] Katta, VR, Takahashi, F, Linteris, GT, "Numerical investigations of CO_2 as fire suppressing agent", In: Evans, DD, ed.,Int. Assoc. for Fire Safety Science, Boston, MA, (2003).

[40] Linteris, GT, Chelliah, HK. Powder-matrix systems for safer handling and storage of suppression agents. NISTIR 6766, Gaithersburg, MD: National Institute of Standards and Technology; 2001.

[41] Linteris, GT, "Suppression of cup-burner diffusion flames by super-effective chemical inhibitors and inert compounds", In: Gann, RG, Whisner, KC, Burgess, SR, Reneke, PA, eds., Papers from 1991-2006 Halon Options Technical Working Conferences (HOTWC), CD-ROM, NIST SP 984-4, HOTWC-2001, National Institute of Standards and Technology, Gaithersburg, MD, (2001).

[42] Linteris, GT, Takahashi, F, Katta, VR. Cup-burner flame extinguishment by CF_3Br and Br_2. Combust Flame 2007; 149:91-103.

[43] Takahashi, F, Katta, V, "Stabilization and suppression of axisymmetric diffusion flames", In: 45th AIAA Aerospace Sciences Meeting and Exhibition, AIAA, New York, (2007).

[44] Goldmeer, JS, T'Ien, JS, Urban, DL. Combustion and extinction of PMMA cylinders during depressurization in low-gravity. Fire Safety Journal 1999; 32:61-88.

[45] Ruff, GA, Hicks, M, Mell, WE, Pettegrew, R, Malcom, A, "CO_2 suppression of PMMA flames in low-gravity", In: 7th International Workshop on Microgravity Combustion and Chemically Reacting Systems, NASA/CP-2003-212376/REV1, NASA, Cleveland, OH, (2003).

[46] Ohlemiller, TJ, Shields, JR, Butler, KM, Collins, B, Seck, M, "Exploring the role of polymer melt viscosity in melt flow and flammability behavior", In: New Developments and Key Market Trends in Flame Retardancy, Proceedings of the Fall Conference of the Fire Retardant Chemicals Association, Fire Retardant Chemicals Association, Lancaster, PA, (2000).

[47] Montegi, T, Shibuya, T, Tsuruda, T, Saito, N. A study on fire suppression phenomena of gaseous extinguishing agents for flammable solids. In: Report of National Research Institute of Fire and Disaster, No. 96. Japan: 2003. 52-57.

[48] Panagiotou, J, Quintiere, JG, "Generalizing flammability of materials", In: Fire science and engineering conference; Interflam 2004, INTERFLAM -PROCEEDINGS-; 10TH 2004; 2004; CONF 10; VOL 2, Interscience Communications, London, (2004).

[49] Grant, CC. Personal Communication. 2008;

[50] Keski-Rahkonen, O, Mangs, J. Electrical ignition sources in nuclear power plants: statistical, modelling and experimental studies. Nuclear Engineering and Design 2002; 213:209-221.

[51] Flamm, J, Bengtson, G, Niemann, R, "Continuing the examination and comparison of existing Halon alternatives in preventing re-ignition on continuously energized fires", In: Gann, RG, Burgess, SR, Whisner, KC, Reneke, PA, eds., Papers from 1991-2006 Halon Options Technical Working Conferences (HOTWC), CD-ROM, NIST SP 984-4, HOTWC-2005, National Institute of Standards and Technology, Gaithersburg, MD, (2006).

[52] Tewarson, A, Khan, MM. Extinguishment of diffusion flames of polymeric materials by Halon 1301. Journal of Fire Sciences 1993; 11:407-420.

[53] Tewarson, A, Khan, MM, "Extinguishment of diffusion flames of polymeric materials by Halon 1301", In: Gann, RG, Burgess, SR, Whisner, KC, Reneke, PA, eds., Papers from 1991-2006 Halon Options Technical Working Conferences (HOTWC), CD-ROM, NIST SP 984-4, HOTWC-1992, National Institute of Standards and Technology, Gaithersburg, MD, (2006).

[54] Niemann, R, Bayless, H, "Update on the evaluation of selected NFPA 2001 agents for suppressing Class "C" energized fires", In: Gann, RG, Whisner, KC, Burgess, SR, Reneke, PA, eds., Papers from 1991-2006 Halon Options Technical Working Conferences (HOTWC), CD-ROM, NIST SP 984-4, HOTWC-1998, National Institute of Standards and Technology, Gaithersburg, MD, (2006).

[55] Bengtson, G, Flamm, J, Niemann, R, "Update on the evaluation of selected NFPA 2001, agents for suppressing class "C" energized fires featuring C_6 F-ketone", In: Gann, RG, Burgess, SR, Whisner, KC, Reneke, PA, eds., Papers from 1991-2006 Halon Options Technical Working Conferences (HOTWC), CD-ROM, NIST SP 984-4, HOTWC-2002, National Institute of Standards and Technology, Gaithersburg, MD, (2006).

[56] Bengtson, G, Niemann, R, "Update in the evaluation of selected NFPA 2001 agents for suppressing class "C" energized fires", In: Gann, RG, Burgess, SR, Whisner, KC, Reneke, PA, eds., Papers from 1991-2006 Halon Options Technical Working Conferences (HOTWC), CD-ROM, NIST SP 984-4, HOTWC-2005, National Institute of Standards and Technology, Gaithersburg, MD, (2006).

[57] Driscoll, MR, Rivers, PE, "Clean extinguishing agents and continuously energized circuits: recent findings", In: Gann, RG, Burgess, SR, Whisner, KC, Reneke, PA, eds., Papers from 1991-2006 Halon Options Technical Working Conferences (HOTWC), CD-ROM, NIST SP 984-4, HOTWC-1997, National Institute of Standards and Technology, Gaithersburg, MD, (2006).

[58] Smith, DM, Rivers, PE, "Effectiveness of clean agents on burning polymeric materials subjected to an external energy source", In: Gann, RG, Burgess, SR, Whisner, KC, Reneke, PA, eds., Papers from 1991-2006 Halon Options Technical Working Conferences (HOTWC), CD-ROM, NIST SP 984-4, HOTWC-1999, National Institute of Standards and Technology, Gaithersburg, MD, (2006).

[59] Smith, S, Gallus, T, Tapia, S, Ball, E, Beeson, H. Electrical arc ignition testing of spacesuit materials. Journal of ASTM International (JAI) 2006; 3:229-246.

[60] Khan, MM. The effectiveness of Halon 1301 in suppressing cable fires ignited by a sustained electrical arc. FMRC J.I. OT1E4.RC, Norwood, MA: Factory Mutual Research Corp.; 1992.

[61] Kurosaki, Y, Ito, A, Chiba, M. Downward flame spread along two vertical, parallel sheets of thin combustible solid. Proceedings of the Combustion Institute 1979; 17:1211-1220.

[62] Siegel, R, Howell, JR. Thermal radiation heat transfer. Washington: Hemisphere Pub. Corp., 1981.

[63] Welty, JR, Wicks, CE, Wilson, RE. Fundamentals of Momentum, Heat and Mass Transfer. New York: John Wiley & Sons; 1976.

[64] Marks, LS. Marks' standard handbook for mechanical engineers. New York: McGraw-Hill; 1978.

[65] Stoliarov, SI, Walters, RN, Lyon, RE. Determination of heats of gasification of polymers using differential scanning calorimetry. 2007;

[66] Twilley, WH, Babrauskas, V. User's guide for the cone calorimeter. SP-745, Gaithersburg, MD: National Institute of Standards and Technology; 1988.

[67] Babrauskas, V, "How do electrical wiring faults lead to structure ignitions?", In: Fire and Materials 2001, 7th International Conference and Exhibition, Proceedings of Fire and Materials 2001 Conference, Interscience Communications Ltd., London, (2001).

[68] Linteris, GT, Rafferty, IP. Flame size, heat release, and smoke points in materials flammability. Fire Safety Journal 2008; 43:442-450.

[69] Linteris, GT, Gewuerz, L, McGrattan, KB, Forney, GP. Modeling Solid Sample Burning with FDS. NISTIR 7178, Gaithersburg MD: National Institute of Standards and Technology; 2004.

[70] Roman, J, Kluge, R, "Key NEBSTM system design and test considerations to minimize TTM and costs", In: Intel Developer Forum, Spring 2003, 2003).

[71] Robin, ML, Rowland, TF, Cisneros, MD, "Fire suppression testing: extinguishment of Class A fires with clean agents", In: Gann, RG, Whisner, KC, Burgess, SR, Reneke, PA, eds., Papers from 1991-2006 Halon Options Technical Working Conferences (HOTWC), CD-ROM, NIST SP 984-4, HOTWC-2001, National Institute of Standards and Technology, Gaithersburg, MD, (2006).

[72] Babushok, VI, Tsang, W. Inhibitor rankings for hydrocarbon combustion. Combust Flame 2000; 123:488-506.

[73] Linteris, GT, Gewuerz, L, McGrattan, KB, Forney, GP, "Modeling Solid Sample Burning", In: Gottuk, DT, Lattimer, BY, eds., Fire Safety Science -- Proceedings of the Eight International Symposium, International Association for Fire Safety Science, Boston, MA, (2005).

[74] Holman, JP. Heat Transfer. New York: McGraw-Hill; 1981.

[75] Mulholland, GW, Janssens, M, Yusa, S, Twilley, W, Babrauskas, V, "The Effect of Oxygen Concentration on CO and Smoke Produced by Flames", In: Cox, G, Langford, B, eds., Fire Safety Science--Proceedings of the Third International Symposium, Elsevier Applied Science, London, (1991).

APPENDIX I – Phone Interviews with Industry Experts on Fire Suppression

Respondent 01

1. Respondent 01 started off by saying that he feels that a major factor in consideration of telecom/data processing fire suppression by clean agents is the massive evolution of equipment which is occurring recently. After the Hinsdale fire, federal agencies were energized, and created the Network Reliability Council. As a result, Bell developed NEBS (Network Equipment Buildings Standard) to specify the requirements for protection of the public network, including fire safety in telecommunications facilities. Consequently, the major producers of equipment (Lucent, Northern Telecom, Siemens, and Ericsson) all had to meet the very comprehensive NEBS standards (one section covers fire resistivity in equipment). For example, one standard specifies a test in which a gas burner is used to start a fire on a rack, and the resulting fire cannot exceed 150 kW power output, and the equipment must be designed so that the fire will not spread across an isle to an adjacent rack. The tests specify such things as the radiant energy, dripping, etc., which can affect the fire spread. Fires in equipment meeting NEBS are very smoky (due to the use of a lot of fire retardant compounds in the raw materials and coatings).

2. Since NEBS implementation, the traditional service providers in the telecom industry have been following NEBS, and it seems to have been very successful. Nonetheless, the Bell system is no-longer the standard in telecom, and the old, standardized system is being overtaken by, new innovative technologies and business models. As telecom service providers move away from NEBS type equipment, the question remains as to how code bodies, such as NFPA 76, will deal with the proliferation of new, varied equipment types, which are not following NEBS. The way NFPA 76 deals with this now is to say that the telecom industry does not have to have a fire suppression system if NEBS is followed; whereas, for typical IBM-type equipment (e.g., bladeservers, typical data centers), a UL standard has to be followed. He does not feel, however, that this standard is as stringent as the old NEBS standard. NFPA 76 produced a table, which required that for the UL standard, a fire suppression system must also be implemented.

3. Respondent 01 then described a report which he will be able to send me in which the types of equipment classes were separated into three levels, and different levels of fire protection were applied to each class. He suggested several contacts for me whom he felt would be helpful (which he provided in a follow-up email). He noted that there are chapters in the NFPA handbook on Telecom and Data Processing Center fire protection.

4. In summary, Respondent 01 described how he feels that the vast and rapid changes in the telecom industry leave it vulnerable to fire threats. It is only a matter of time before there is another major fire disaster in telecom. The building standards have not kept up with pace of the changes, both in the equipment types and servicing approaches. The NEBS standards were good, and companies followed them, but many new, innovative start-up companies do not follow them.

5. Respondent 01 understands that Telcordia, the "owner" of the NEBS standard, is developing a comparable standard for "Non-NEBS"/non-traditional electronic equipment, but he has no direct knowledge of its content or status.

Respondent 02

1. Fire Examples:

a.) They have had some fires in their facilities.
b.) They have been in the rectifiers and in batteries.
c.) In Illinois: they had a fire that cracked a battery, leaked material onto steel, shorted it out, created a fire, the spill container bags went on fire, heated jars, failed and put out the fire. There was lots of smoke.
d.) In another fire, a rectifier failed in the middle of night, and the suppression system (Halon) put it out. In their industry, they still have quite a bit of Halon systems left; they have not been replaced, for example, in long distance sites. They also have other clean-agent systems.
e.) They have had no cable fires.
f.) They had a fire before the Halon system was activated. There were manuals on top of a carrier, which got so hot it caught on fire.

2. System details:

a.) They want to prevent events.
b.) They have very early warning systems in a lot of the systems, so they stop a lot of fires.
c.) Most of their web equipment is NEBS compliant.
d.) On the network, if it comes to a fire alarm, they will discharge someone.

Fire protection is based on:
1. very early warning smoke detection,
2: regular smoke detection,
3. suppression: (Halon or HFC),
4. in some locations, sprinklers.

Based on previous experience, the philosophy of their employees is to never shut the place down. But, on a number of occasions, they have shut down (e.g. floods), but they never have shut down for fires.

Some carriers have disconnect switches. But now, with homeland security concerns, there is a mandate for "no single point of failure." So what do you do?

There is one carrier who has water over their system. In the switch room where they have sprinklers, they have pre-action sprinklers (which requires two detection zones to charge pipe). In their system, they must have smoke or heat, and then it charges the line with water; then if there is enough heat, the water releases.

They have a power down procedure, that is being modified, and they currently have a draft of a practice for power disconnect. The procedure will be posted on equipment.

Respondent 03

There is a range of feelings on the NFPA committee on Fire Suppression of Electrically-energized Class C fires: some feel that for electrically energized fires, there is a need for higher concentrations than the 2001 Class A numbers, while others do not.

He has witnessed tests of electrically energized fires with clean agents (either CF_3Br or some of the new agents): the agent suppresses the fire, and then when the flow of agent is shut down, the fire re-flashes.

In his industry (Telecom), the Central Office facilities (at least where the equipment is) generally are not protected with sprinklers. There is no suppression in the central offices. The reason is that with either Halon or sprinkler, Class C fires will keep on cooking until the power is turned off. So there is a power-off procedure to use, followed by addition of suppressant. In contrast, in Information Technology offices, there is often an emergency power-off button. He and others say that this may be counter productive, since this approach has now introduced a single point failure possibility which would not otherwise exist.

As for the clean agents, he feels that adding more agent probably still won't put out an electrically-energized fire, but will make the room more dangerous for people. It's better to lose equipment than to lose people.

Datacom power supplies are 120 V A/C, whereas telecom are 48 V DC.

Over the years, very few fires have occurred in telecom, and no-one has died in a telecom fire. The fires which they have had have been in power cables as opposed to data cables. The power cables are big, thick cables, and can have a short.

In the computer world, there are not a whole lot of fires in Datacom. His peers in Datacom tell him that they do not consider the threat very significant. Article 645 of the National Electrical Code say that for Information Technology equipment room, there should be an EPO (Emergency Power Off) switch.

Telecom has always been very careful with respect to fire safety. Central offices are exempt from any suppression requirement. This is likely the result of extensive fire detection equipment, and the implementation of the NEBS. Other rooms in the facilities (e.g., cafeterias, work spaces, etc, must have fire suppression systems installed). Some of the older buildings in the network of one telecom company have Halon or sprinklers. Another telecom is different: they don't do NEBS, but they do sprinkler.

He felt that many in the telecom industry feel that water is more of a threat than fire. Some switches will start a fire if they get water on them. The damage in some telecom equipment is different than for computers; in telecom, water can have a very detrimental effect. An additional consideration is that telecom has a "don't lose service" requirement from the government.

He says that his company has no situations that he knows of in which the procedures are to release a clean agent into a Class C.

Data centers: lots of equipment is not NEBS compliant, so their solution is just to sprinkler it. (The spaces are too big for gas systems.)

In NEBS tests, they test the equipment at the frame level, shelf level and circuit level. (The circuit pack is 25.4 cm by 15.2 cm by 2.54 cm wide; they slide on runners to a back plane, that the circuit pack plugs into.) Each level cannot pass a flame from one to the other. The NEBS standard goes way beyond fire, to include electrical and physical failure also; e.g., earthquakes.

None of the Bell companies will put equipment into a central office if it is not NEBS compliant, and they do not co-mingle NEBS and non-NEBS equipment. The NEBS standard goes way beyond the relevant datacom standard (UL 60950).

FM did tests back in 1999. A real live switch, powered up, and had a whole bunch of scenarios, with sprinklers putting out the fires.

He says that most of telecom's fires are cold and smoky, and not really big.

To summarize, he says that they have no scenario in which they will release a clean agent into an energized electrical component. Also, they really get all NEBS compliant equipment into their central offices.

Respondent 04

Respondent 04 believes that, while many Class C fires will be put out by clean agents at the Class A level, there exists some concern that some class C fires will not be put out. Hence, the 2001 committee may not be doing the right thing by keeping the Class A numbers to apply to Class C fires.

In the real world of operators of data centers, computer rooms, telecom, he estimates that less than 20% would turn off the power before activating a clean agent fire suppression system.

While he thinks that there are some who understanding that the Class A numbers may not be correct for some Class C fires, he feels that in the marketplace, most do not understand that the Class A number might sometimes not be high enough; they only know that 2001 says to use the Class A numbers.

This respondent felt that, realistically, if more agent is required, it's bad for the clean-agent fire suppressant system industry—but that does not mean that we should not do the right thing and try to understand what the real numbers are. Maybe it's still ok if we find, and can understand, that most fires will go out, but not all.

As for the detrimental effect on industry, part of the problem may be that if the required concentrations get too high, then release of the agents may be unsafe for the occupants, and hence, the systems will be less attractive. The owners might decide that they may as well go back to water. Another effect is one of cost: in a large data center, these systems can be very expensive, 10^6 or 10^7 dollars; with 5 000 to 10 000 lbs of agent, the cost of the agent itself can be 50 % t o 75 % of the total cost.

Respondent 04 suggested that perhaps the idea for Class A fires is that the UL number is some sort of "worst-case scenario"; but is it really? Perhaps this is how we should go for Class C fires: come up with a worst case situation and use that as a design. The respondent and I discussed that another approach, and one which may be sensible, it to use the power input as a parameter. That is, the required design concentration for the Class C fires will likely depend upon the amount of power added to the flammable system. There is a very wide range of possible power levels for Class C fires. One could then specify the amount of agent required for a Class C fire according to the amount of added electrical energy. It would be up to the system designer to decide what the amount of energy added to the flammable materials might be.

Respondent 05

Respondent 05 provided a very long and detailed commentary, lasting several hours, which included: three case studies, a short list of other fire events, and general comments. Discussion focused on telecom Central Offices, but covered other areas as well.

Case Study 1

1998 UPS system

In the uninterruptible power supply (UPS), there were 240 cells, nominal 540 VDC Valve Regulated Lead Acid (VRLA) batteries. They had an immobilized electrolyte in a polypropylene jar inserted into a square metal box. A pressure regulating valve prevents off-gassing during charging, and the gases recombine, thus avoiding the need to add water the battery. The VRLA battery system had been moved from another location and the polypropylene (PP) jar had somehow been damaged. During the subsequent use of the battery, the internal pressure buildup ruptured the weakened battery jar and electrolyte leaked out at two locations. The epoxy-coated battery rack was eventually damaged and thus a path to ground was created; the flow of current was sufficient to create massive energy release. Current flow could approach short circuit current of the ≈ 1500 A · h (8 hour rated batteries). Current flow was high enough to cause the PP to burn with a flame, as well as well as to cause conductor failure, since there was no over current protection. The auto-start smoke exhaust system in the building was able to control the smoke and heat so the automatic sprinkler system did not activate in the ≈ 929 m^2 (10 000 ft^2) battery or power-plant room. The exhaust of the smoke prevented serious equipment contamination. The good-house keeping practices in the building meant that ordinary combustibles were not present. Hence this electrical fire, while likely very dramatic and violent with molten metal being ejected, and 7 pounds of metal being vaporized, was not sustained long enough to cause any major flaming combustion. Engineers estimate that this was approximately two ≈ 400 kW electrical events, with combustion heat release of an additional 50 kW. The discharge was rather short lived, and self-terminated with conductor failure, as determined afterwards (since the batteries still had the majority of the stored full charge). Smoke exhaust not only provided control of containments from the smoke, and removal of heat, but provided visibility for the responding fire fighters. The emergency personnel were reluctant to enter the space since they saw warning signs for components containing PCB. Had the room been protected with a clean-agent system, contamination would have been worse since the smoke exhaust system would not have been run; this would have, in the opinion of Respondent 05, resulted in a worse situation, with more contamination, and greater required cleanup.

Case Study 2

UPS Charger

Later that afternoon a manager was investigating the fire in the UPS. As the manager was taking notes he smelled a 'new overheated electronics smell' which would come and go. He discussed this with the technicians on site who were attempting to put the UPS back on line with a small string of 'car batteries' which would have had 2 min of reserve. Eventually they traced the problem to the UPS charger, and opened up the charger to check for hot spots with an IR camera. The technician was looking into unit from the top, through a cooling fan exhaust. The metal oxide varistor (MOV) housed in the UPS exploded, most likely because of extended AC over-voltage and/or phase imbalance which caused excess MOV heating. The technician was burned by the jet of flame and 'fire works' that resulted. An employee present got the technician off the UPS charger and placed him in a chair and wheeled him to the elevator. An ambulance met the employee and the technician in the lobby when they arrived there. Smoke detection did eventually respond but not before the employee and the injured technician got to the lobby. As a result of this incident, the company subsequently got into VEWFD.

As another example of a battery fire, there was a fire in a lead acid battery bank in California. Technicians were removing end cells, and replacing them with a special set of cells that would be inserted into a discharging / depleted battery string, to boost voltage and prevent an imminent network outage (the design is no longer being used). The activity apparently went well and several battery strings were corrected when, from what Respondent 05 knows, the battery voltage dropped and then, one the end cell systems was 'deployed' for the first time. Unfortunately the voltage was reversed. This caused major conductor damage due to high current flow and grounding. A typical design uses 500 and 750 MCM cable, so current flows are huge (even though only 48 VDC). Current flow caused the battery jar to burn, but it was not a major fire. The fire fighters apparently walked past the seat of the actual fire during their response, because the heat release was relatively minor, and the flames were shrouded in intense smoke (there was a lot of smoke damage). The electrical heat release and current flows were large relative to combustion heat release. The fire was put out with a 'pail of water' and then manual smoke exhaust was started.

Case Study 3

Bus-Bar Accident

About a year later, 1999, a major electrical event. In a telecom Central Office (CO), workers were putting in two new diesel backup generators and a new switchboard. On Saturday night, the final connections for a new switch board were made during a blackout. On Friday morning: workers were installing the grounding bus bar in the main breaker cell next to the 13 800/600/347 V 3 phase transformer main disconnect bus of the building. The workers dropped the new bus, which hit a cardboard box, which was not enough to protect and insulate the live busses, and it shorted. The power was 4000 kW; an oak ladder 1.5 m (5 ft) away got charred due to radiation heat transfer. Double interlock sprinklers came on, and 75 7000 L (20 000 gallons) of water discharged (which meant that near the ceiling, the temperature was about 811 °C, or 1000 deg F). The Central Office had to de-power, but there was lots a lot resistance from people to do this. There was a worker on the floor of the room with the bus bar (he lived). The main problem was that there was no rapid de-powering scheme to be used during electrical emergency situations.

Other fire experiences.

1. Burn out of a 10 hp return air fan motor. Fire detected everywhere in building given the air circulation from the fan.

2. Short of a 600 amp 240 V, 3 phase power supply – mechanical interlock device disconnected but was left in the switchboard.

3. Shorting of power conductors to composite sheet fire-stopping material, causing fire.

4. Power plant short. 2 500 A 48 V DC power plant line was grounded during cable placement. Note: the cables were 750 MCM with no over-current protection. Current flow at least 50 000 amps. Arcing lasted about 4 min until 16 individual, 750 MCM cables were cut by technicians doing installation work. Took about 4 min, and the fire department was on the scene when they came out of building. Arc fault potential and safety risk.

5. Cable connector failure in cable tray. Arcing and overheating. Power technician called fire department and then depowered DC power plant. Fire self extinguished.

6. Transformer fire – prolonged commercial power failure August 14th 2003. Poor crimped conductor between dual windings of the transformer arced and burned out 6" section before gap too large to sustain arc. Approximately 100 kW event.

7. Several building fires due to overheated diesel exhaust stacks, some in diesel generators, and other fires due to construction and hot work.

8. In general more water damage than fire losses.

9. There is no average telecom electrical fire scenario; it depends on where the fire occurs. Size varies significantly.

10. To reduce the risk: a) use good housekeeping practices to limit presence of ordinary combustibles; b.) use equipment and wiring meeting high fire resistance standards; and c.) use de-powering schemes when possible.

General Comments:

1. Central Office Buildings are very complicated with respect to electrical power.
a.) AC phone power distribution
b.) AC essential power
c.) AC non-essential power
d.) DC power plant with duel feeds to every load
e.) unfused DC batteries string to buss connections
f.) UPS power.

2. An electrically-energized fire will cook until you turn of the power. Decomposed materials will be produced without flaming.

3. But, if you have a suppression system, at which point in the process do you release the clean agent gas? Very early warning fire detectors can detect fires an hour or so before a regular fire detector would. So when do you release the agent? Too early, and the ignition source will still be there, and it will just relight after you have lost the suppressant.

4. Inert gases have the advantage for clean agent suppression in that they have no breakdown products. Very important for Electrically-energized fires (EEF).

5. Most DC conductors in telecom COs are way over sized (based on voltage drop, and they want to minimize this). So the amount of energy in a EEF can be huge in the sections of the system without over-current protection, such as the battery string. There, the power is the higher of the available amps or the failure current of wire, times the voltage.

6. There are many parts of the system which are unfused (batteries are there for backup, and this DC power source is typically unfused, with NO central disconnect).

7. A particular telecom has had many little events, failed components, etc., which blow holes in the boards, but since all the equipment is NEBS, you don't get a fire. The materials have strict FR guidelines. The respondent recalled a test in which FE36 was being tested, tried to get some circuit boards going, but they wouldn't burn enough to get something to put out. They used the NEBS ignition test, but even then flames are only a couple of inches long and the fire spread is not even over the entire 8" x 12" circuit board. Adjacent boards have even less material consumed by fire.

8. If using a hand-held fire extinguisher, the hard part is to get the agent to the right part of the equipment where the fire is, and then add your agent at the right rate so it
    a.) gets in to reach the fire; i.e., the agent stream does not deflect, and
    b.) does not get wasted (by using too high an addition rate).

9. For NEBS equipment, one fire risk is:
    a.) packaging material- used for shipping, when new equipment arrives, can be all over the place (cardboard, foam liner, etc.) . If this were around when there was an ignition event, the problems would be much worse, a real fire and likely fire spread into the electrical system, as wiring is everywhere.

10. The range of power dissipated in EEF:
    a.) a few hundred watts, to
    b.) massive amounts of energy (e.g., 4000 kW in bus-bar or battery accident).

11. A very good approach for EEFs is to use manually released clean agents when possible, released right where the fire is.

12. In telecom, everything is supposed to be NEBS compliant. But, purchasing agents can—in trying to get the lowest price, buy the wrong thing; e.g., cabling. All kinds of cable can end up in a cable tray, including non-fire retarded. Centralized purchasing by people who don't know the difference in the products, can accidentally lead to the wrong products being used. The telecom manufactures are very good at making products with the right rating. This respondent has observed problems with power cable and non metallic routing assemblies; many different commercial applications exist and these have varying fire resistance.

13. There is more and more outsourcing, with little oversight, which can create more variability, more permissiveness in what is used and how the work is performed.

14. Fire retardants can be driven off over time if the electrical short is slowly cooking for a while.

15. In cable trays, the wires age, shift, and deteriorate, and are exposed to vibrating equipment hangers etc. These can make the fire risk higher than when the equipment was new.

16. In Telecom, must consider foreign voltages, which can make things behave very differently.

17. A fire accident in telecom is just a matter of time.
    a.) telecommunications central offices following NEBS are safe.
    b.) the early warning detection systems are very effective.
       but... ..
          i.) the buildings are not always staffed, and
          ii.) the staff present do not always know what to do. They may not have been trained on the fire procedures, or on de-powering procedures.

18. The situation in NFPA 75 is more tenuous: Some of the equipment and materials are the same as telecom NEBS in terms of the materials used but the final assembly does not have to pass the same strenuous fire safety and limited fire performance test, and many of the power cables are permitted to have lower fire resistance. Note NFPA 75 assumes that fire suppression will be present, whereas telecom has fire prevention designed into the permanent network equipment and wiring. The telecom situation is one of excellent fire prevention by an integrated performance-based system; in the IT world, there is more reliance on fire suppression. These environments are being co-mingled with the convergence of IT and telecommunications. But if there are DC power circuits, these may be unfused within the DC part of the UPS, and many of the wire types may be non-fire retarded. One could have a problem. In general, can't extrapolate what we know from one, familiar situation (NEBS telecom) to another which is new (e.g. data processing). However every major telecom fire disaster has been electrical in origin or it became an electrical fire due to lack of depowering. There have only been a very limited number of such major fires, and with no loss of life. This is over a 100 plus years and some 200M sq. ft. of space in North America.

Respondent 06

Clean agents are heavily used in: data centers, switching centers, UPS, and transformer rooms.

In electrical equipment, fuel sources include pc boards in a cabinet, wire, bundles of wires, cables, etc.

In efforts to date, there really have not been people who have come forward and offered up, as examples, actual fires which have occurred. Hence, Respondent 06 thinks the subject of suppression of electrically-energized fires is characterized by a little bit of knowledge.

There exists the understanding that if you have an arc, there is a continuous ignition source, but if there is continuous ohmic heating, the hot surface can also provide ignition. Clearly, if there is a continuing ignition source, it affects how you put out the fire.

There exist two lines of thought:

1. If you have a local ignition, it burns up the material in the locality—but after that, there is no fuel left nearby. So as the burn zone moves away, there is no more electrical augmentation to the fire.

2. The other perspective is that if you have a fire with ohmically heated cabling, (e.g., a fire in a cable bundle), it has it's own fuel source, and now a secondary heat source. In this case, the ohmic heating makes the primary extinction of the fire more difficult, and beyond that, if the extinguishing agent discharges, and then dissipates, but the power is still on, then the fire might re-ignite.

Trying to define the problem is has been the problem.

The end-user community, by insisting that they will maintain power, is inviting the challenge.

A lot of the challenge is in coming up with an appropriate test method is because the actual electrically-energized fire to be extinguished is hard to define.

So what is the path forward? One path is to identify one specific hazard, then identify the right test to characterize the suppressant requirement for that threat (i.e., identify the effect of electrical energy on the suppressant concentration required for extinguishment), and then move on to another, then another, etc. For example, UPS room and batteries. One would have to talk to a Fire Protection Engineer who understands what the actual fire risks are in this case, discerned from close calls, failures, etc. Do it here, and then move on, on a case by case basis. We need a specific example of how having an EEF affects a specific case. How about just one single case, that meets the core criterion of concern, and then see how does the inclusion of added electrical energy affects the failure mode—how does added power affect the suppression?

Respondent 07

This respondent said that he is not as active now at the front lines of fire suppression systems as he was before.

The big question that must be addressed is: What is the risk?

He suggested that the risk in a UPS, battery room will be of one type, but in a data processing room / switching room, the risk will be very different.

Respondent 07 feels that the big telecom companies have been approached to provide information on energized electrical fires. But what about the service people? Smaller people in the field may have some very good information, and may be able to speak more freely. They might have information on fires and agent discharges that could be helpful.

In the old days, CO_2 was the clean agent. But the standard for CO_2 was based on different power levels (higher) in the equipment, and different, more flammable materials (not fire retarded). He feels that the CO_2 standard as it now exists is probably too conservative for modern situations. He feels that clean agent standards need to be applied only for the cases for which they were developed. For example, clean agents in a data processing equipment room have a very different fire threat than in rotating electrical equipment.

Respondent 07 stressed that the datacom equipment industry represents a very broad range of equipment types, power levels, and fire threats; energized electrical equipment in general will be even more varied. Thus, in this project, we must characterizes what we are describing, so that the standard developed for one equipment and fire threat is not applied in the wrong place.

Respondent 07 is sure there is a fire risk in energized electrical fires, but he is not sure to what extent.

For clean agents, Respondent 07 guesses that 90% of the industry is in telecom, and data processing. Industrial users are small fraction of the total. Data/telecom uses include: telecom: switch rooms, datacom anywhere and everywhere; power generation/industrial facilities: main control rooms.

Respondent 07 said that in the old days, they had trained personnel with very detailed fire suppression procedures. As an example, he cited a control room in an industrial facility that had a Halon system backed up by a CO_2 system. They had a very detailed procedure for a fire alarm, and the staff was well trained. That sort of level of procedure is not so common in the industries typically protected by clean agents today.

Respondent 07 mentioned that some reports have described how of the fires in telecom, most have been of other flammable materials, not the electrical equipment itself.

Respondent 08

No further input is necessary.

Respondent 09

He and colleagues have visited places. The big threat that people appear to be concerned about is power rooms: big room with all the batteries, main power, etc. Sometimes these are different rooms from the switching rooms, sometimes the same room. The battery rooms can be capable of 5000 A, with one inch diameter copper cables. The batteries can also create harmful gases which people are also concerned about.

One challenge in defining the fire threat in energized electrical fires is that there have not been many fires. Most of the fires which have occurred have involved the power rooms. There have also been small, low-power fires of boards: a component failure, smoke, fire, etc.

In general, power supplies in racks seem to be an issue. Some companies are selling lots of small systems to extinguish fires in cabinets (mostly for power equipment and safety switches). He has heard that there are around 50 fires per year in these type cabinets (in aggregate, for a number of cites totaling about 10 million people).

He feels that it is generally hard to know where the actual electrical fires are.

In tests at one company, they used (18 to 22) gage wire, and drove them to failure. To do larger wires will be difficult.

Respondent 10

From a computer electronics standpoint, there have been very few actual fires from computer equipment itself. He only knows of two, and they were small:

1. Around 1998, a power supply failed and smoked before the circuit protection kicked in. The smoke set off the suppression system; no flames.

2. Around 1983, in some premesis wiring there was a short in the equipment power plug/receptacle, which lead to smoke in the underfloor cabling area, setting off an alarm; no flames. The equipment had a flexible power cord (plastic or rubber depending on ampacity, etc.) with the mating plug. This plugged into the customer's branch circuit, which was a flexible power cable like BX type cable with the appropriate power receptacle and fittings attached. This was located under the raised floor within a few feet of the equipment to which it supplied power. In this particular case, one of the receptacle branch conductor connection screws was not tight. Over time, this created an overheating condition and the presumption was that smoke came from either the actual branch phase conductor or the thermoplastic receptacle / plug housing. It may not have ever been determined which, since the heat traveled from the receptacle to the plug, but the receptacle had the loose connection, everything had that bluish-white smoke residue, and the conductor insulation was charred.

There have been cases of fire suppression discharges. These are usually related to adjacent room fires in which enough smoke came into adjacent room, to set off alarm.

There was a report by NFPA, around 2000, which talked about electronic fires. They did not classify them as data center vs. other equipment. He will see if he can find the reference.

The type of equipment that he deals with is in data centers that fit the article 645: power disconnect, fire rated material, separate HVAC rooms, etc. The data center rooms vary in size from 27.9 m^2 (300 ft^2) to 9 300 m^2 (100 000 ft^2). He has a lot of familiarity with data centers all over the country and world.

Clean agents are an insurance policy, The NFPA policy is: if your building has sprinklers, so does your data center… …but water and data centers don't mix too well, hence the attraction of clean agents.

As far as he knows, no one has really done a test with fire in a live piece of equipment to try to put it out with a clean agent; it could be very expensive. This is true of any of the agents, including hi-fog. There are not a lot of data available on suppressing energized electrical fires. A consideration is that with a clean agent, one only gets one shot to put the fire out (since after that, the agent is depleted). Of course, the first response to a fire will be for an on-site employee to use a handheld extinguisher. Downtime is no good.

The National Electrical Code allows for small UPS (less than 750 W), so if you lose main power, can still power some equipment. So some rooms have many of these present. Insurance companies have also argued to eliminate any central disconnects, since this is a site of one-step "failure" (e.g., disgruntled employees, mistakes, could easily bring down the whole system, and bypass all the built-in redundancies).

The interviewer brought up the point that others have said that sometimes what goes on in a data center is not well controlled, since it is not specified by a large, single decision company like in the old days. Hence, there could be more variability in terms of materials used and conditions.

Responder 10 replied that because of requirements of the data center owners, they are very strict about who is doing what in the data center at what time. Because of constant changes, there is a lot of activity in a data center (lots of changes). But the activity is very tightly controlled, because they don't want to lose service. Some of the customers (i.e., data center owners) are even more strict than large companies. They have done so much planning, in term of redundancy (for example, what they can take offline, when, etc.), that they are very careful about being sure that no-one in there is going to bring them down by accident.

There has been a lot of innovation in terms of power. More and more IT equipment is built with dual power cord capability, so work can be done without losing equipment. Another goal is to develop smokeless systems, so if you have a failed component, it will not make any smoke.

He would have to presume that in a data center, for any fire, the source of ignition would be electrical. From a data center view point, no one has combustibles sitting around. They are very clean; some places you can't even bring in boxes. There has been a lot of pushback from clients: no boxes in the rooms, and this is becoming more and more common. Hence, one large vendor thought about eliminating packaging, but the replacements weren't very good. Some data centers forbid the packaging material, so the vendor asked for staging areas, which they got. At other data centers, they said you can bring in boxes, but at the end of the day they must be removed.

Ignition sources? There's lots of equipment 208 V, 40 A to 50 A.

Things have changed a lot in the past thirty-five years. The old equipment had 100 A power plugs, and a data center might have been 46.5 m^2 (500 ft^2). Today, typical centers are in the 465 m^2 (5000 ft^2) to 930 m^2 (10 000 ft^2) range; about 5% are 4650 m^2 (50 000 ft^2).

There have been a lot of changes in the standards for power connections, with regard to flammability ratings, number on connection points, etc.

In the end, Responder 10 agreed that it would be nice to characterize how much clean agent is required to put out an electrically energized fire, but that the problem is very broad, because of the very wide range on types of equipment and possible fires which would need to be considered.

Respondent 11

Carriers are putting fiber closer to the home.

CO have lots of battery juice.

One company went with lithium polymer batteries out in the field. They had three of these explode, leading to battery-caused fires.

In one case, in 17 000 locations: the green box at the end of the street, 1m (3 ft) by 1.5 m (5 ft) tall, a major company has been putting extra electronics in them. They specified a big fat cable to go into these, UL94? rated, but then, later discovered that the cables that actually made it in were not the proper fire rating. After discovering the error, they had to take 17 000 of these out.

If you go into a data center, most of the equipment meets some UL, or FCC standard ok, but other than that, there are no real standards, so they are moving in the direction of having everything comply with some more universal standard. They have their own standard now, that is an amalgamation of a lot of other standards, so for now, they at least have something.

Respondent 12

NEBS has tests for fire spread on the chassis. As part of that, they put a line burner in the chassis, and look for evidence of self extinguishment or spreading. There is a limit on the peak, and average heat release rate of the fire.

The test runs about 6.5 min, then stop. The test is modeled after a burning circuit pack, of a certain size. This respondent suggested that in our EEF tests, we could use this size fire as a the level of the heat source for the energy augmentation. At least it gets us in the ball park in terms of the amount of energy added to the system. The idea is that in NEBS, if this amount of energy is added to the circuit pack (from an external source, which in this case, is a fire, not an electrical source), the fire must not propagate for the system to pass. In our case, with that amount of added energy (simulating an electrical source), the clean agent must be able to extinguish the *other* materials burning in the cabinet.

This respondent feels that a lot of the data center equipment is pretty close to NEBS, and that there is a lot of movement to get these close.

Respondent 13

In the nuclear industry, as with large turbine generators, there are safety concerns which stipulate that one can't just shut the equipment off (e.g., in turbines, there are backup DC lube oil pumps, that provide lube to bearings while systems turns down).

From previous experience, he knows of computer rooms, where the procedure was to trip out the computers before releasing Halon.

Basically have electrical and electronic rooms, cable spreading rooms, that all have control functions to them. In nuclear power plants, can't turn off the switch; have to maintain cooling functions.

Nuclear plants run off emergency diesel generators; don't use batteries too much.

Years ago they used CO_2 in their industry. Sometimes they use water, but for the most part, it's all detection based.

Respondent 14

The problem of electrically-energized fires can be separated into two basic areas: 1.) low voltage, low energy; e.g., data centers and telecom; and 2.) power plants. He questions the applicability of clean agents in heavy industrial applications.

This respondent has had a lot of interface with telecom. They do not want to de-power prior to release of a clean agent; whereas, in NFPA 75, with its facilities, they do recommend de-powering—but recognize that it is often impractical to do so. (For example, if a data center is monitoring a hospital with ICU telemetry, would not want to de-power.) Still, it is preferred to power down.

Sometimes a room will be a legacy system (e.g., 1301), but there will have been many changes to the equipment configuration over many years, so that the original design parameters have not really kept up with the room changes.

In datacenters, there are people running around, so you don't want to use CO_2, and you don't want to use water. So clean agents have found a niche, but how much agent do we need?

Maybe over the years, the problem of added electrical energy has been sort of ignored. Probably also the case with Halon too, but that agent was so much more forgiving; we have a smaller window to work with, so it may be more important to consider.

Respondent 14 thinks that in an ideal world, building owners/operators would follow NFPA 75, not have combustibles in the room, use fire-retarded plastics in housings, have cables with fire-retarded insulation, etc. Historically, the problem in NFPA 75 has been in getting history: it has been difficult to get data on fire events. But this might be good news: it may not be a big problem.

In the end, Respondent 14 agreed that a reasonable approach is to explore the parameter space, see how much more agent you need with electrical energized fire, and let the user decide on how much agent is necessary based on the fire type, material, and electrical energy input.

Respondent 15

Suggestions on how to bound the problem:

Low energy, low power applications seem to be where the action.

To keep things bounded, probably not trying to protect high energy cable fires.

Instead, look at protecting 1.5 kW of power or less. An individual piece of equipment (i.e., a cabinet) is usually 1.5 kW or less so, might want to make this the cutoff.

While both DC and AC power are important, might want to keep these two problems separate.

Respondent 16

He does not know what the threat is in energy-augmented fires. He has not heard of any problems in the field.

He feels that the big thing with this project is: Is there a threat? and if so, what exactly is it?
He agrees that in practice, defining the treat is difficult.

He says that under NEBS, enclosures are tested for fire resistance, and as a result, the fire risk may be lower than it used to be. So we have fire resistance, but also have fires suppression systems.

He felt that getting data from the field on exactly what the fire threats are would be very useful, but that it might be difficult to get at the information. He suggested that the telecom/datacom industry might try to do what the Dept. of Homeland Security has done, in setting up a near-miss database for first responders. They can provide, anonymously, information on events. They contribute since ultimately, it is in their interest to be safer. But the question remains as to whom in telecom/datacom would benefit if they contribute.

He suggested that I look at the recent presentations at SUPDET to see what recommendations have been made with regard to test procedures for suppression of energy –augmented fires. Newer recommendations may be different from previous recommendations.

We discussed how various UL and ANSI tests are the basis for NEBS equipment, and may provide a starting point for test development with energy augmentation. He suggested that
NEC National Electrical Code requirements would serve as a starting point as well.

Respondent 17

I described my job as to: 1.) figure out what the fire threats are which people want to suppress with clean agents, and 2.) define a test procedure that would be used to determine the amount of agent needed to suppress an energy-augmented fire.

Respondent 17 said that the reason why they sent this project to NFPA Research Foundation—to define what the threats are. He believes that the problem is very large and loosely defined: from small electrical equipment to power plants. There are many possibilities. The biggest issue is defining the hazard.

As far as clean agent applications are involved, most of the protection (90 % to 95 %) is in telecom and data processing. He does not think that high voltage, high current have to be considered at all. He feels that the committee agrees: low voltage, low amperage is where to look.

He feels that a real survey of the industry is needed to define what the real threats to be protected are.

Respondent 18

1. With electrical involvement, one is not really worried about energy augmentation, because it all happens so fast, and then the ignition source if off. But on the other hand,

2. With electrical ignition, one gets a very fast ignition, over a much larger area, so suddenly, it's acting like a much larger fire, with heat feedback, so the usual flammability vs. the radiant flux arguments apply.

3. Cable fires have shown this very fast, large ignition area with very fast spread and initial heat release.

4. Look at Sandia/NRC cable fire studies.

5. Hinsdale is not representative, since the damage from the fire was mostly from pyrolysis products; i.e., smoke damage, not burning.

6. With suppression, one wants to catch the fire very early, otherwise clean agents does not work. But, with electrical ignition, get a much bigger fire early on, so it *will* be harder to put out, because of heat feedback to burning material.

Respondent 19

Talked about characterizing the fire threat for electrically-energized fire (EEF) suppression.

1. Respondent 19 said that he felt it was a non-issue. First question: is it an arc-flash fire or not? If the electrical source is less than 240 V, it's usually not an arc-flash fire[4]. For less than 240 V, any arc will be short-lived, will go out soon, and does not sustain itself. There is not really much energy from the source adding to the fire.

If more than 240 V, (480 V and up), can have, for example, a run of arcing bus bars, which can produce a ferocious energy release. In this case, no automatic system will put it out, and no fire chief would add a suppressant to this.

2. He thinks the problem is misunderstood in the NFPA literature.

3. There was a famous Atlanta explosion related to electrical shorting, at 25 Peachtree Plaza. There's a FEMA report on it. It started as a humongous arc-flash. But even here, the electrician would not have been helped by a clean agent. The arc fire ignited a corridor, flashed over, so whatever the fire load from the adjacent flames were, this would be the type of environment of the burning material. If you look at the conditions of the Atlanta vestibule, you might say that a room corner fire test is the most similar standard test to simulate the conditions.

4. Respondent 19 agrees with Respondent 18's feeling that with electrical ignition, the area of the initial fire involvement, and the subsequent faster growth rate would make electrically-ignited fires generally more difficult to put out than other fires without electrical involvement.

5. On a related topic, Respondent 19 feels that the he NFPA handbook section on purported dangers of energized electrical fires on fire-fighters is good, but often is not read by fire chiefs. As a result, fires are often not attacked because the power company has not yet turned off the power. He thinks this is not the course of action which the data suggest should be taken.

The most recent work on this is a SNITEF (Norway) report by Vierlo (≈1996). In it, they describe the best equipment for an electrically-energized fire, and try to quantify the safe conditions for extinguishing an electrical fire. For example, with fog nozzles, one can get to within 2", and with a hose stream (which should not be directed at a switchboard), you could get to within a few feet. With 480 V, don't want to get anywhere near it with a hose.

The point is that sometimes, electrical fires are not fought, but often they ought to be.

_____

[4] I am not sure I understand this comment. As indicated in the work of Tewarson and Khan [53,60], and McKenna et al.[4], clearly a very low voltage DC source (on the order of 10 V to 30 V) can cause a continuous arc.

APPENDIX II – Case Studies Supplied by FM Global

Case Study 1 (Page 1 of 2)

Year of Incident: 2006

Hazard Summary

Electrical overheating is one of the leading ignition sources in the industry. In this case, overheating of a group of plastic insulated electric wires, inside a static power switch, resulted in heavy smoke development. Although the thermal damage was limited to the electrical cabinet of fire origin, due to its light combustible contents and the manual fire-fighting efforts by both employees and the Public Fire Department, there was extensive non-thermal smoke damage to the data processing equipment and building itself.

Site Specific Conditions

The facility is dedicated to the leasing of space and utilities for Electronic Data Centers (EDC). The Main Building is concrete on protected steel and has two stories. It is subdivided by 2 hour fire-rated partitions into 16 rooms (8 in each level) of 400 m^2 to 500 m^2 each. (Refer to attached sketch.)

The room involved was occupied by lines of cabinets containing servers dedicated to xx clients. It also contained three or four data handling computers. The room has an uninterrupted power supply (UPS) fed, normally, by the 11 kV public supply and three on-line dynamic UPS diesel-driven generator sets. Both these supplies feed two medium voltage systems (A & B) which, in turn, feed the data processing rooms at 380 VAC. Inside the leased rooms, these main feeds are interconnected through a dynamic switch. This switch is an electronically controlled high-speed device which, when needed, automatically transfers the feed to the room, under load, between the A and the B systems. It consists of a metal cabinet (approx. 1 x 0.8 x 0.5 m in size) containing ventilated bus bars and an electronic switching command circuit. This is manufactured by ABC company, and it is reportedly rated at 380 V – 160 A.

Reportedly, no open flame work or soldering is allowed within EDC rooms.

Protection of all data processing rooms is by a clean agent total flooding systems which is activated by automatic smoke detection. Furthermore, suitable numbers of carbon dioxide fire extinguishers are available throughout. Alarms are transmitted to the permanently attended guardhouse in the lobby of the Office Building. Interlocks reportedly automatically trigger the Emergency Power Off system (EPO) in the room. Reportedly, a delay of 30 s has been set between the smoke detection and the clean agent system discharge to allow for room evacuation or resetting in case of false alarm.

The Public Fire Department (PFD) servicing the area is located in xxxxx. An advanced first intervention station is located only a few blocks away in the industrial park. Both stations are permanently manned with well-equipped, professional fire fighters. The site has been visited and intervention plans established by the PFD.

The utilities systems both supplying and within the rooms are reportedly adequately maintained and tested, in accordance with good standard practices, by the owner of the site buildings and utility equipment. The data processing equipment is owned, installed and maintained by renter's staff.

The Incident

The following information is reported the facility manager of the leased space Electronic Data Center:

Workers from a subcontractor of the property owner were pulling cables under the raised floor of the room to install a sixth static switch. They reported what they call three explosions in sequence within one of the five static switches in the EDP room located on the top floor of the Main Building. During work in the room, the clean agent extinguishing system was placed in manual operating mode. This is due to concerns with a false trip of the system during such work and where the operators might not reach and operate the discharge interrupt switch in time to prevent discharge.

It is believed that the EPO system interlock is designed to normally operate only upon flow in a room's clean agent system which shuts down the room's power supply and ventilation systems. The incident is followed by heavy smoke development from within the switch's metal enclosure. At about the same time, the guards in the office building receive a smoke alarm from the detection system and immediately summon the public fire department. In the mean time, employees go to the Medium voltage room in the room directly below the room of fire origin and electrically isolate the electrical breaker switch feeding the room above. Employees then proceed to open the cabinet and discharge CO_2 extinguishers inside the cabinet. Somewhat later following their arrival, the public fire department fully extinguished the fire and starts ventilating the room. Upon inspection, the fire's thermal damage was confined to a section of about 4 in. (10 cm) of a grouped low-current, plastic-insulated cables inside the metal cabinet's enclosure. As soon as the extinguishing tasks were completed and declared finished, the site's catastrophe recovery plan was activated and later on, the clean agent protection system was placed in automatic service.

Damage Summary

According to the facility manager of the leased space, the equipment in the entire room was contaminated to various degrees by the smoke. Some smoke contamination is also suspected in the room to the East via the corridor along the North wall of the building. This contamination will is currently being reviewed by three independent forensic companies.

A decision has been made by local lease management to replace all equipment in the room due to suspected unreliability.

It was noted, at the time of this visit, that all the equipment relocated from the fire incident room to the adjacent room appeared to be operating normally. At the time of this visit, all equipment had been inspected and cleaned. The extent of damage could not therefore be ascertained.

Cause

The ignition source of the fire could not be established. The cause is still under investigation by independent forensics and the equipment will ultimately be reviewed by the equipment manufacturer.

The most likely main cause is an uncleared electrical fault in the metal electrical cabinet of fire origin.

Case Study 2   (Page 1 of 2)

Year of Incident: 1997

Hazard Summary

This high-rise building used as office, telephone central office, and computer center was visited to investigate a fire.

A fire occurred in an (manufacturer's name) data storage array. Detection by ceiling mounted smoke detection in conjunction with fast response by property personnel and the Public Fire Department helped contain and limit the damage. The building has 11 above grade stories and 3 below grade. Building construction is reinforced concrete floors and a combination of concrete block and gypsum board interior walls. Part of the 6th floor is utilized as the Operations Center. The Operations Center consists of a 10 000 sq. ft. (100 x 100 ft.) east computer room and an 8,000 sq. ft. (80 x 100 ft.) west computer room. The east and west computer rooms are separated by the Operations Center central control center which is occupied 24 hours a day.

The remainder of the 6th floor contains a mechanical room, conference room, storage room, and a network equipment room. The telephone switching equipment used for high speed transmission of data for business clients.

The west computer room has a 0.3 m (1 ft.) high raised floor and a mineral tile drop ceiling 2.7 m (9 ft.) above the raised floor. The computer room is separated from other occupancies by gypsum board walls. The wall between the west computer room and the room is concrete block. No protection is provided in the room. The room is equipped with smoke detectors. These are analog detectors which are capable of reporting response information and location. The detectors were originally installed with a 1.2% sensitivity and are spaced every 37.2 m² (400 ft²). Detectors are located below the raised floor, at the drop ceiling level, and above the drop ceiling. These detectors are monitored by the Operations control center and the building maintenance central control center. The computer room has 4 kg (9 lb.) Halon 1211 extinguishers placed throughout. There are four personnel doorways located near each corner of the room. As you exit each, an emergency power off switch is provided.

The west computer room has 7 Liebert units used for humidity and temperature control within the computer room. 'These Liebert units are interlocked to shut down upon activation of smoke detection within the room. The west computer room is only 50% occupied with equipment at this time. The room contains various computer complexes which run various software functions.

The data storage memory array is approximately 1.82 m (6 ft.) long x 1.1 m (3.5 ft.) deep x 1.82 m (6 ft.) high. The unit is separated into three bays with each bay being identical on the front and back. The middle bay contains the ac input module. AC current enters the input module at 208 three phase, Delta configuration. Current then runs through one of the three ac to dc transformers which steps down the current to either 5 or 12 V. Current is then run on four busses around the outside of each bay for delivery to the disk drives. System requirements are 5 V at 500 A and 12 V at 290 A. The central bay also contains four 35 A · hr. emergency batteries which will provide current for up to 3 min. to assure no data loss during power outages. The final components of the middle bay are 20 control circuit boards. Each end bay contains 5-1/4 in.

hard drives. There are eight rows of four hard drives each from top to bottom. The hard drives are 5-1/4 inch high x 4 in. wide x 9 in. long. Each hard drive is connected to a midplane which runs the entire width of the bay, separating the front and back of the unit. Each hard drive is encased in plastic. Plastic materials within the data storage array are reported to be polycarbonate, fire retardant polycarbonate, and PVC. Bays are separated by sheet metal. On top of each bay are fans which exhaust heat out of the cabinet. The
end bays have six fans and the middle bay has eight fans.

INCIDENT

At xx:xx a smoke alarm activated in the west computer room operations center. This smoke alarm automatically notified the computer control room, the building control room, sounded the audible building alarm, and shut down all Liebert units. At approximately +4 min, the computer room operations center control room attendant opened the door to the computer room and found thick black smoke. (All doors were closed prior to the arrival of the attendant.) He immediately called 911 to verify the fire. While awaiting for the fire department, the Operation's Center attendant gathered as many Halon 1211 extinguishers that he could find from the floor. At +14 min, the fire department arrived and entered the room. They were unable to find the fire due to thick black smoke. Visibility was reported to be less than 1 m (3 ft.). At +22 min, the fire department activated the emergency power off switch. The fire department attempted to ventilate the room by opening doors from the computer room. Doors to the network equipment room were also opened as the fire department attempted to find a way of venting the smoke outside of the building.

During this time, it is reported that smoke migrated outside of the Operations Center to other areas of the 6th floor, including the network equipment room. At +24 min, the fire department was able to extinguish the fire with several of the Halon 1211 extinguishers. Mechanical ventilation of the room was conducted 1 hour 45 min from the start of the incident, when the building was released to the owner.

Owner's personnel began disaster recovery procedures 3 hours after the start of the incident. The computer room was mopped and equipment wiped down to remove smoke residue. Ceiling tiles in the immediate area of the fire were replaced.

CAUSE

Approximately a 6 x 6 in. (15 x 15 cm) area of the midplane in the disk storage array was consumed. The plastic casing on approximately 10 of the hard disks was partially or totally consumed by fire. Heat within the disk array was not hot enough to melt any of the copper wiring, as it was found intact.

Evaluations are currently being conducted regarding the extent of non-thermal damage to the computer equipment within the Operations Center. The owner has conducted 150 wipe samples of the equipment and is working to evaluate the extent of non-thermal damage.

Year of Incident: 1993

A fire occurred on an automatic voltage regulator in the Electronic Data Processing Center. It was put out by the automatic Halon systems (twelve, 64.5 kg (142 lb.) cylinders discharged in total). The EDP center is on the ground floor of the Office Building and is approx 400 m^2 (4300 sq. ft). It is enclosed by masonry walls and/or REI 60 partitions, gypsum board incombustible ceiling and chip wood floor, coated on the underside with an 0.8 mm steel sheet. EDP center core is the VAX room, 170 m^2 (1800 sq ft) approximately, containing three Digital 8500 VAX, a Digital 8800 VAX, eighteen RA81, 454Mb disk units, twelve RA82,622Mb units, four RA72, 1 Gb units, other CPUs and all the modem units for internal data transmission.

The plant operates on a highly integrated system and this room manages all the production.

The EDP center is protected by four, total flooding extinguishing Halon 1301 systems in: EDP ambient, EDP underfloor, VAX room ambient, VAX room underfloor. The activation is controlled by a smoke detection system. The system was installed in 1990.

At 00.00 hours, the shift VVF in the guard house was alerted by the fire alarm panel of "extinguishment on" in the EDP centre.

The four firemen in the guard house reached the spot, joined by a fifth fireman who was performing his round. On arrival, they soon realized that VAX ambient, VAX underfloor and EDP underfloor systems were on. Not seeing any flames, they entered the VAX room to check that the fire was totally put out by the Halon (in total, twelve 64.5 kg (142 lb) cylinders discharged). Noting that the fire occurred in a IREM, 45 kV, 380 V voltage regulator (type 1/32/48, reg. No. 243844) on a perimetral wall of the VAX room, that the Halon had completely put out the fire and that the electrical protection on the general feeder were on, de-energizing the whole room, they opened the doors and windows to evacuate the extinguishing agent (+2 min). The Fire Chief then authorized the maintenance technicians (informed by the guard house according to the Emergency Plan procedures), to begin the repairs in order to put the system back in service.

At about +45 min the various EDP center supervisors and the Fire Officer arrived.

At +75 min a bypass on the burnt regulator was installed and the bootstrap operations on the systems began. .

At +150 min the on-line command was given to activate the production lines, and at +165 min the usual, first shift production started.

At +105 min the smoke detection/Halon extinguishment system was put back in service by Firemen, using the back-up cylinders. During this visit the Fire Officer said that the empty cylinders would be sent to be recharged.

RESULTS

The voltage regulator is completely destroyed.

The sudden de-energization of the electronic equipment, operating at the time of the incident, did not cause any damages to the disk units or software. After the needed restarting operations, production could be restarted without any problem.

CAUSE

According to electronic laboratory personnel, the voltage regulator could have caught fire because of an overheating automatic regulation rheostat, which never activates as the regulator is downstream the UPSs of the EDP centre that regulates the power.

CONCLUSIONS

The Halon protection extinguished the fire on the voltage regulator in the VAX room before it propagated to the underfloor cables and then to the other CPU apparatus. This evidence (fire spreading to underfloor and then to other Vax equipment) is proved by the cables, located on the rear of the voltage regulator, partially burnt and evidently put out by the Halon system.

In case the fire had happened during a working period, there would have been a total shutdown of the EDP center for three-four hours. It should be remembered however that the smoke may have been noted by operators in this case.

The installed protection operated correctly and the recommendations contained in this report will reduce the probability of a reoccurrence.